Department of the Environment
Geological and Minerals Planning Research Programme

REVIEW OF THE EFFECTIVENESS OF RESTORATION CONDITIONS FOR MINERAL WORKINGS AND THE NEED FOR BONDS

Michael Whitbread
Christopher Tunnell

Arup Economics & Planning

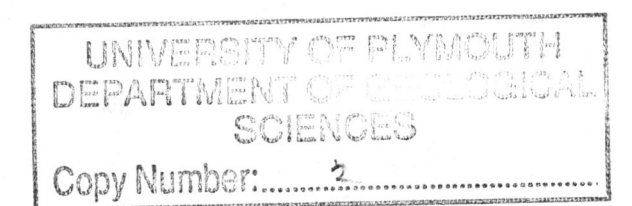

London: HMSO

© Crown copyright 1993
Applications for reproduction should be made to HMSO
First published 1993

ISBN 0 11 752764 5

Acknowledgements

Front cover pictures:

Sand and gravel working and restoration of a site in the Wheeler Valley, North Wales. Photos reproduced with permission of Bodfari Quarries and the Sand and Gravel Association.

Review of the Effectiveness of Restoration Conditions
for Minerals Workings and the Need for Bonds

CONTENTS

Page

Preface

EXECUTIVE SUMMARY i

1 INTRODUCTION
    1.1 Aims of the Study 1
    1.2 Research Methods 2
    1.3 Earlier Studies 3
    1.4 An Outline of Existing Planning Procedures for Restoration Following Mineral Workings 7

2 CURRENT PRACTICE AND EXPERIENCE REGARDING MINERALS RESTORATION 11
    2.1 Surveys of Local Authorities and Mineral Operators 11
    2.2 Assessment of the Effectiveness of Existing Procedures 29

3 EXPERIENCE OF FINANCIAL GUARANTEE SCHEMES 32
    3.1 Voluntary Agreements 32
    3.2 South Wales Local Acts 35
    3.3 Industry Schemes 37
    3.4 Ironstone Restoration Fund 39
    3.5 Indemnities, Guarantees and Performance Bonds under Contracts 40
    3.6 Overseas Experience 43
    3.7 Principal Findings from the Experience with Financial Guarantee Schemes 50

4 ALTERNATIVE FINANCIAL GUARANTEE MECHANISMS 52
    4.1 Restoration Liabilities 52
    4.2 Purpose of the Financial Guarantee 52
    4.3 Decentralised Bond and Guarantee Instruments 54
    4.4 National Scheme : Restoration Fund 60
    4.5 National Scheme : Insurance Fund 61
    4.6 Conclusions on Alternative Mechanisms 64

5 CONCLUSIONS AND RECOMMENDATIONS 66
    5.1 Limitations of Existing Procedures 66
    5.2 Scope for Improvement to Planning Practice 68
    5.3 Policy on Financial Guarantees and Bonds 70
    5.4 Schemes for Financial Guarantees and Bonds 72
    5.5 Summary of Recommendations 74

APPENDICES

1   Postal Questionnaires of Planning Authorities and Mineral Operators
2   Respondents to the Postal Surveys
3   Participants in the Follow-Up Interview Survey

LIST OF TABLES

		Page
Table 1.1	Area of surface mineral workings with no provisions for restoration and aftercare	4
Table 1.2	Causes of major failure to observe satisfactory restoration conditions	5
Table 1.3	Restoration performance 1974-1982	6
Table 2.1	Response rates for local authority postal survey	12
Table 2.2	Restoration performance 1982-1990	13
Table 2.3	Restoration performance by type of operator	14
Table 2.4	Reasons for unsatisfactory or no restoration and/or aftercare	15
Table 2.5	Level of enforcement action in relation to sites identified as having unsatisfactory or no restoration	22
Table 2.6	Regional distribution by type of operator	23
Table 2.7	Number of operator sites ceased or completed operations 1982-1990	23
Table 2.8	Reasons stated by operators for unsatisfactory restoration	25
Table 2.9	Responsibility for ensuring that planning conditions for restoration and aftercare are carried out	25
Table 2.10	Policy on financial provision for ensuring compliance with site restoration and aftercare conditions	27
Table 2.11	Summary of reasons for failure	30

Table 3.1	Restoration bonds held by planning authorities under Voluntary Agreements or other arrangements	33
Table 4.1	Annual restoration costs by main mineral type (Great Britain)	53
Table 4.2	Estimated costs of a bonding scheme for current mineral sites	57
Table 4.3	Costs of a National Insurance Fund	64

LIST OF BOXES

Box 1	Example restoration conditions taken from a recent permission for a quarry	16
Box 2	Example restoration conditions for short term operations on an opencast coal site	19
Box 3	Example of a financial guarantee by Agreement under the planning acts	34
Box 4	Extract from a South Wales local act	36
Box 5	Example bond wording - South Wales	37
Box 6	Specimen form of on demand bank guarantee	40
Box 7	Example of insurance company bond	42

Preface

This report was commissioned by the Department of the Environment in March, 1991 from Arup Economics & Planning. The majority of the work was carried out by Michael Whitbread and Christopher Tunnell, drawing on technical expertise from within the Ove Arup Partnership in relation to geotechnics, the environment, contracts and legal aspects. Arup were supported in their research by a DOE nominated Steering Group whose guidance and technical inputs are gratefully acknowledged. The Steering Group comprised:

Mr Robin Mabey	(Chair) Department of the Environment
Miss Ann Ward	Department of the Environment
Mr Michael Burn	Department of the Environment
Mr William Watts	Department of the Environment
Dr Tom Simpson	(Secretary) Department of the Environment
Mr Brian Spiers	Scottish Office Environment Department
Mr G R Jones	Welsh Office
Mr Roger Caisley	Derbyshire County Council
Dr Michael Gandy	Mid Glamorgan County Council
Mr John Stewart	Central Regional Council
Mr Bryan Frost	Ready Mixed Concrete Ltd
Mr Keith Bramley	British Coal Opencast Executive
Mr Ron Parry	Tarmac Quarry Products
Mr W H Ward	Country Landowners Association

The research would not have been possible without the willing co-operation of many people and organisations who were interviewed or otherwise provided information, including the completion of postal questionnaires. Details are provided in Appendix 2 and 3. Every effort has been made to accurately reflect the comments which were received during the course of the study. However, the findings and conclusions reflect the views of the study team alone and do not necessarily represent the views of the Department of the Environment or any other parties involved.

EXECUTIVE SUMMARY

Background

1. The Stevens Committee Report, **Planning Control over Mineral Working** (HMSO, 1976) examined the operation of planning controls over mineral workings and recommended amendments to provisions. Amongst the issues considered by Stevens was the need for a system of financial guarantees or bonds to ensure that if a planning authority was unable to secure compliance with conditions for restoration and aftercare of mineral extraction sites through established enforcement remedies, the necessary funds could be made available for the works to be carried out.

2. Stevens concluded that only those cases of restoration failure which could not be dealt with by the proper application of current or amended Planning Law and procedure should be considered for a financial guarantee or bonding system. Stevens considered that the cases most likely to fall into this category derived directly from the financial or similar failure of the operator. However, Stevens did not recommend the introduction of a new system but suggested the matter should be kept for further review.

3. Currently planning authorities in Great Britain (other than three in South Wales with Local Act powers) cannot impose a financial guarantee or bond as a condition of planning permission. Over the past few years, however, an increasing number of bonds have been provided in Agreements with mineral operators relating to restoration and aftercare following mineral extraction. About one hundred of these Agreements exist in Great Britain and although the practice appears to be increasing experience of these Agreements is limited and does not provide the basis for a proper assessment of the need for new instruments nor how a comprehensive financial guarantee or bonding system might work. Other experience with bonding arrangements both in Britain and overseas provides some additional but still only partial evidence and insights. A wide-ranging study was therefore required for the matter to be reviewed.

4. The Department of the Environment commissioned Arup Economics & Planning in March 1991 to undertake a study as part of Government's commitment, in response to the findings of the Stevens Committee, to keep under review the effectiveness of restoration conditions applying to minerals planning permissions. The study has examined whether additional measures such as financial guarantees or restoration bonds could achieve better or more rapid restoration results in practice.

5. The specific aims of the study were:

    - to examine the extent to which planning conditions requiring restoration of mineral workings in Great Britain are achieving their aims within the development control framework;

- to update the Stevens Committee's assessment of the need for a performance bond or levy system to secure the restoration of mineral workings, taking account of the current situation and experience in Great Britain;

- to outline the preferred form of alternative best options which any financial guarantee system should take if the Department were to consider that some statutory provision should be introduced;

- to provide a report for Government as a result of the work, which takes account of the views of the minerals industry and mineral planning authorities.

Study Tasks

6   In order to achieve these study aims several separate pieces of work were undertaken. Extensive postal questionnaires were sent to all planning authorities in England, Wales and Scotland with minerals responsibilities, and to a large number of mineral operators. Response rates were high for a survey of this kind, being 58% overall for the planning authorities and 35% for operators.

7   To supplement the postal survey findings, personal interviews were undertaken with 20 planning authorities and 20 operators to obtain insights into attitudes and opinions, and to explore in greater depth their restoration experiences.

8   The survey results were compared with those obtained from earlier surveys undertaken by Stevens, the County Planning Officers Society in 1982/83, and the two DoE surveys for 1982 and 1988, Surveys of Land for Mineral Workings in England.

9   Investigations were made of the experiences of various financial guarantee schemes. These covered:

- the limited experience of financial guarantee and bonding arrangements in Britain under voluntary Agreements associated with mineral planning permissions;

- experience of operating with powers under Local Acts which have been given to the authorities of West Glamorgan, Mid Glamorgan and Dyfed in South Wales, in relation to financial guarantees for restoration after private coal mining operations;

- mineral industry guarantee schemes, particularly that of the Sand and Gravel Association;

- the Ironstone Restoration Fund which ceased operations in 1985;

- contract indemnities, guarantees and performance bonds such as those used for highways and construction projects;

- overseas experience.

10 Personal interviews were held with several financial institutions, both with respect to their experience of operating bonds for a variety of different purposes and for insights into the ways that the banking and insurance industry might approach the issue of financial guarantees and bonds for mineral restoration and aftercare.

11 Using the evidence collected in these study tasks, the Consultants identified a number of possible financial guarantee and bonding schemes which could be applied to the processes of mineral extraction in Great Britain. These were evaluated according to three criteria: their likely effectiveness in achieving planning objectives; their impact on existing enforcement procedures, and the costs imposed on industry.

Findings

12 According to planning authorities, about 27% of minerals sites ceasing operations between 1982-90 were not restored to satisfactory standards or were not restored at all. This proportion has remained broadly the same since the time of Stevens although it is certain that standards being used in the judgements of what is satisfactory have risen.

13 However, the mix of reasons for this failure to restore has changed. The survey of mineral planning authorities identified three principal sets of reasons for failure to restore mineral sites to satisfactory standards, and a variety of lesser reasons:

- financial or similar failure of the operator, which has occurred at or near the exhaustion of the mineral extraction process for the site in question. As was found in Stevens' survey, this represents only a small proportion of failure cases, at 5%;

- technical failure by the operator, for which there was a variety of possible types and reasons. This category represents about 42% of failure cases, greatly up from the time of Stevens;

- failures of implementation of the planning system itself, such as badly worded conditions, lack of appropriate monitoring by the planning authority and failure to enforce against breaches of conditions. This represents about 35% of failure cases, well down from the time of Stevens;

- a balance of 18% of failure cases included ownership and other operator problems.

14 The mineral operators surveyed did not think that failure to restore sites for whatever reason was as prevalent as the extent considered by the mineral planning authorities.

15 With respect to the failures of implementation of the planning system, evidence suggests that local authorities have made significant strides, leading to improvement in overall performance. Local authority efforts have also been supported by legislative changes in the 1981 Minerals Act. It is anticipated that the 1991 Planning and Compensation Act will lead to still further improvements.

16 The study has concluded that powers that are now available largely meet the requirements of the planning system to achieve its objectives of restoration and aftercare but certain problems of implementation do remain:

- resources, both in terms of staff and finance, appear inadequate for many authorities to undertake all the tasks required of them;

- modernisation of old permissions, in particular, is not proceeding rapidly enough and this remains a major cause of concern;

- some authorities are reluctant to take enforcement action against defaulting operators due to delays, uncertainties and costs;

- all authorities express difficulties in pursuit of successors in title when failure results from liquidation of the operator;

- there are some special difficulties associated with restoration of borrow pits used for highway and similar projects.

17 The other principal category of failure to restore, technical failure by operators, remains a significant problem and is 7-8 times more prevalent (according to planning authorities) than financial failure. Technical failure is the failure to comply with restoration conditions for which there is an explanation to be found in the adverse physical circumstances of the site or restoration materials or the poor practices adopted by the operator. Standards are increasing and it is difficult to disentangle the extent to which technical failure remains a serious problem because of enhanced expectations and the extent to which some operators lack the technical capability or incentive to undertake a good restoration job. Clear differences of attitudes and opinions were found between the planning authorities and the operators about failure to restore, with operators claiming far lower incidence of unsatisfactory restoration.

18 Circumstances most likely to lead to successful restoration include: adequate site investigations prior to working; planning permissions with realistic, clear and comprehensive conditions; presentation to authorities by operators of a well worked-out scheme of restoration and afteruse linked to financial and material budgets; technical competence by the operator both in relation to mineral workings and restoration; proper monitoring of activities by the authorities, and a willingness by the operator to do a good job as evidenced, for example, by environmental audits. In many instances the operator has an incentive to secure or maintain a good local or national reputation so that further permissions to work new or expanded sites may be obtained more readily.

Possible Approaches to Financial Guarantees and Bonds

19 There is no direct overseas experience to provide a model of the operation of financial guarantees and bonds for restoration which can be applied directly to the British planning system. British experiences with existing voluntary Agreements and the South Wales Local Acts are both limited since in neither case have guarantees or bonds been called other than for operator financial failure (and only then in the South Wales cases), and only a few have been called for this reason. Close consideration of these bonding arrangements suggests, in fact, that they have value as a security against financial failure only, and that technical failure is not covered because of difficulties authorities would face in demonstrating to the satisfaction of the courts that default had occurred.

20 In approaching the issue of financial guarantees or bonding arrangements for mineral restoration, therefore, it is necessary to be clear about the objectives of the scheme. If a new scheme is to be a comprehensive financial guarantee against technical as well as financial failure by the operator the scheme needs to ensure that technical failure is capable of being clearly demonstrated:

- either, there needs to be a clearly identified responsibility for the determination of whether a breach of conditions has occurred, with only limited appeal procedures; or

- conditions need to contain highly specified schemes of restoration broadly similar in the degree of detail to those encountered in contracts such that technical breach of conditions can be clearly identified and demonstrated.

21 Either of these approaches would mark a radical departure from existing British minerals planning procedures and, if introduced, in the view of the Consultants would amount to a replacement to the existing enforcement procedures. Few authorities would prefer to engage in lengthy, costly and uncertain enforcement action if they can have ready access to funds to undertake the works.

22 A more limited alternative to a comprehensive financial guarantee or bonding arrangement would be one which is only intended to cover financial and related default. Occurrences of financial failure are clear-cut. They offer bondsmen no difficulties for deciding when to release funds. They would meet some concerns of planning authorities particularly about the pursuit of successors in title. But few default cases arise from financial failure and the costs of introducing a scheme would need to be weighed against the limited objectives that would be secured.

Schemes for Financial Guarantees and Bonds

23 Financial guarantee and bonding arrangements fall into two principal groups:

- **Decentralised** - where the planning authority and operator agree to an appropriate instrument of guarantee. Maximum flexibility is maintained in decentralised arrangements for operators to seek their most cost effective solution in the circumstances prevailing, subject to satisfying the authorities' requirements.

- **National** - where a scheme would be comprehensive in coverage and would probably need to be initiated by Government. It could be operated by an Agency or could be contracted out to the private sector to operate.

24 Under the decentralised scheme, mineral operators would simply be required to satisfy the planning authority and would do so in any one of a number of different ways. These include bonds that would be obtained from banks or insurance companies, cash deposits, title deeds and other forms of security which would be lodged with the authority. There would be scope

for mutual fund schemes where groups of operators may come together to provide a collective security (such as the SAGA Fund). Individual guarantees tend to be more expensive for the smaller operators to obtain and individual bonds and financial guarantees on all minerals sites would probably cost industry as a whole between £21-63 millions p.a.

25 A national scheme could be either, a restoration fund (such as the Ironstone Restoration Fund) which assumes financial responsibility for all or part of the restoration function or, an insurance fund which provides for the risk of default only. A national scheme would require that all operators became members since only this requirement would confer the benefits of security for the local authorities. In exchange for membership which would become, in effect, a license to operate, the mineral operators would need to pay a fee or levy to cover the fund's costs.

26 Stevens was not in favour of a restoration fund approach, pointing out that this would take restoration responsibilities away from operators which would be undesirable, and would entail an unnecessary administrative cost. The Consultants concur with Stevens on this matter. An insurance fund approach would assume the risks of default only and this would need to be defined to cover either, financial failure alone for which the national fund costs would be about £3 millions p.a. or, financial and technical failure in a comprehensive scheme for which the fund costs would be about £19 millions p.a. Thus, a national insurance scheme would be substantially cheaper than a decentralised scheme, certainly in the formative years until risks were clearly established. Lower costs arise from the advantages of insuring "a book". It seems likely that larger operators would be obliged to cross-subsidise smaller ones if the levy was imposed without discrimination between different sizes and other types of operator since the smaller operators, as a group, are perceived as a greater risk.

27 There are several alternative approaches to how a levy could be imposed, but its size would probably need to vary by type of mineral and by proposed afteruse, as do restoration costs. A levy could be based on a number of possible measures which are relatively easy to obtain including site area or site output measured by tonnage, volume or value.

Summary of Conclusions and Recommendations in Relation to Financial Guarantees and Bonds

28 Clearly, the issues involved in introducing financial guarantees and bonds for mineral restoration are complex and the arguments for and against different arrangements are subtle yet potentially far-reaching in their implications for the established planning system. Under some of the alternative schemes, the costs of guarantees imposed on industry would be significant and any change to existing procedures needs to be carefully considered. In order to provide a focused basis for discussion and consultation of the policy implications of introducing a restoration bonds scheme and other amendments to restoration procedures, the Consultants offer the following recommendations.

29 The Consultants' principal conclusions and recommendations in relation to financial guarantees and bonds for mineral restoration and aftercare are as follows:

- The Consultants are not in favour of a new scheme of financial guarantees or bonds. However, the basis of many existing voluntary Agreements, which are not common at present but which are likely to increase in the future, is not satisfactory. Future Agreements need to be put onto a consistent basis, in order to avoid discriminatory, costly and essentially irrelevant bonding arrangements coming into effect. This may be achieved by the issue of Minerals Planning Guidance as to the circumstances under which it would be appropriate for a planning authority to reach an Agreement with an operator in relation to financial guarantees, forms of guarantee, checks that the local authority might make on the financial institutions offering guarantees and related matters.

- The Consultants believe that improvements to updating older mineral permissions together with the recent changes to existing enforcement powers are sufficient to achieve restoration objectives in most cases.

- However, if a comprehensive financial guarantee or bonding scheme to cover both financial and technical default was judged to be necessary, then the Consultants recommend that the appropriate route to take would be to keep the existing system of planning consents and to permit financial guarantees and bonds to become conditions of permissions. Legislation would be required. It would be necessary to satisfy bondsmen and others as to the circumstances governing the release of bonds in some cases of default. Most probably the easiest arrangement to introduce would be arbitration procedures to ensure that the release of the financial guarantees and bonds is fair.

- Alternatively, Government could introduce a new scheme which is less ambitious which would provide a partial rather than comprehensive system of financial guarantees and bonds. This limited scheme would be for the purpose of providing security only against the financial or similar failure of an operator to carry out restoration and aftercare following mineral extraction. This partial financial guarantee and bonding system would supplement existing Planning procedures. The limited benefits of guaranteeing against financial failure would need to be weighed against the costs of establishing a new scheme.

- Choices about the approach to guarantees would have to be made in relation to a new scheme, if it is to be introduced. The Consultants believe that a national fund scheme has the advantage over a decentralised scheme, being substantially cheaper for the industry overall and that a national insurance fund would be the best approach. A restoration fund approach is not favoured by the Consultants.

- A national insurance scheme would need to be initiated by Government and would probably need to be operated by a Government Agency although there are other possible options including contracting out. All operators would have to be members of the insurance fund in order to qualify to operate. The levy payable to the fund would need to cover the fund's restoration liabilities, administration and other costs. The levy could be tied to the firms' outputs.

Further Recommendations

30   Other recommendations as a consequence of undertaking this study, are as follows:

- The Study has not found that failure to make adequate financial provision by operators is a cause of restoration failure. Nevertheless, the tax system for companies acts as a disincentive to make provision in some cases. Government may wish to reconsider tax rules in relation to provision for restoration.

- The possible extent of compensation payments has caused difficulties for authorities wishing to modernise permissions. There is a case for extending IDO procedures to sites that have no restoration conditions at present.

- Borrow pits present authorities with particular enforcement problems. Performance bonding for restoration of borrow pits under the terms of the construction contract might be the best route to ensure compliance in these cases.

1 INTRODUCTION

1.1 Aims of the Study

An objective of planning control over mineral workings as stated in Minerals Planning Guidance Note 1 (MPG1) is:

> "To ensure that land taken for mineral operations is reclaimed at the earliest opportunity and is capable of an acceptable use after working has come to an end".

This objective is normally achieved through restoration and aftercare conditions of planning permissions for mineral developments. These conditions are implemented when mineral workings or phases of workings cease. Where an operator fails to implement restoration and aftercare conditions, the planning system provides for remedial enforcement action.

Not all mineral sites are reclaimed to satisfactory standards and this study considers the need for additional instruments to bring about satisfactory mineral restoration. It has been carried out as part of the Government's commitment, in its response to the findings of the Stevens Committee, **Planning Control over Mineral Workings** (HMSO, 1976) to keep under review the effectiveness of restoration conditions and to examine whether additional measures could achieve better or more rapid results in practice. Specifically, the study considers whether the introduction of restoration bonds or levies would be appropriate. The study forms part of the work which the Minerals Division of the Department of the Environment is undertaking to review the operation of the Town and Country Planning (Minerals) Act 1981.

The specific terms of reference for this study were:

- to examine the extent to which planning conditions requiring restoration of mineral working in Great Britain are achieving their aims within the development control framework;

- to update the Stevens Committee's assessment of the need for a performance bond or levy system to secure the restoration of mineral workings, taking account of the current situation and experience in Great Britain;

- to outline the preferred form or alternative best options which any financial guarantee system should take if the Department were to consider that some statutory provision should be introduced; and

- to provide a report for Government as a result of the work, which takes account of the views of the minerals industry and mineral planning authorities.

The report considers the success of existing development control mechanisms in achieving restoration objectives taking account of different sensitivities produced by the extraction of the main mineral types. The report reviews the case and options for financial guarantee or bonding arrangements to supplement or replace existing mechanisms.

1.2 Research Methods

In order to achieve the study objectives the key tasks for the research were:

- to review the practical implementation of restoration and aftercare arrangements under current planning legislation and to identify the extent of and principal reasons for default;

- to examine the scope for additional measures and, in particular, to review comprehensively the options for a financial guarantee system and assess levels of risk and costs.

Data Collection

Data collection undertaken in respect of the study objectives has included:

- desk research, including a literature review of restoration experience and performance;

- postal surveys of all planning authorities in England, Wales and Scotland responsible for minerals and a large sample of minerals operators, with the surveys including requests for data on the success/acceptability of current performance on restoration and the scale and rate of restoration failure and reasons for default;

- focused follow-up interview surveys with planning authorities and operators, covering perceptions about how well the current mechanisms for restoration are working and the problems faced in practice;

- interviews with a sample of financial institutions covering advice and experience of operating bond and other financial guarantee arrangements and collection of evidence of procedures, indicative financial costs, and approaches to risk;

- a review of relevant experiences of financial guarantees, funds and bonding systems, both in Britain and overseas.

Analysis and Report Structure

Chapter 2 of this report includes sets of tables and statistics derived from the surveys which provide data on restoration performance, the scale of and reasons for the failure to restore to satisfactory standards, enforcement and other actions taken in such cases, and matters relating to practical problems of restoration. The current survey results are compared wherever possible with findings from previous surveys. Additionally, Chapter 2 presents opinions of planning authorities and operators concerning the adequacy and limitations of existing policy and practice.

The survey of financial institutions and review of experiences were intended to assist with identifying possibly financial guarantee mechanisms that could be introduced to supplement or replace existing enforcement instruments. This evidence is assembled in Chapter 3. The survey of financial institutions is not reported verbatim since the discussions which were held were unstructured and free-ranging in their nature. But the specific experiences of current and previous financial guarantee mechanisms are described in full, with reports on a number of overseas examples.

Chapter 4 brings the survey findings together to review the scale and nature of restoration problems that could be addressed by a system of financial guarantees and to identify realistic alternative mechanisms. These options have been evaluated according to the likely effectiveness of achieving restoration objectives, their costs of implementation and the potential impact they may have on existing procedures.

Chapter 5 presents the study conclusions and recommendations.

1.3 Earlier Studies

The Stevens Committee

The Stevens Committee was appointed in 1972 in response to general concern about mineral extraction and despoilation. The Committee's remit was to examine the operation of planning controls over mineral workings and to consider whether the then existing provisions needed to be amended or supplemented.

The Committee recommended a separate minerals planning regime with its own provisions for minerals applications and mineral permissions and that planning authorities should have powers to regulate existing workings and require the industry, at its own cost, to meet environmental standards. The Government rejected the notion of a special regime, but many of the Committee's other recommendations were incorporated into the Town and Country Planning (Minerals) Act, 1981.

Some of Stevens witnesses argued in support of bonds or other forms of financial guarantee to ensure that if the planning authority for good and sufficient reason was unable to secure compliance with the restoration conditions through enforcement remedies, the necessary funds could be made available to ensure that restoration nevertheless can be carried out.

Stevens concluded, however, that only those cases of restoration failure which could not be dealt with by the proper application of amended planning law and procedure should be subject to a bonding system. Stevens found that those cases most likely to fall into this category derived directly from the financial failure of the operator in circumstances where the mineral site concerned has no further economic life. These cases of financial failure were relatively limited in occurrence. Stevens hesitated to reject bonding altogether but suggested the matter should be kept for further review.

Table 1.1 summarises data on restoration conditions from Stevens and compares the findings with the more recent Surveys of Land for Mineral Workings in England carried out by the Department of the Environment in 1982 and 1988. Table 1.1 indicates that overall the proportion of sites having no restoration conditions has declined from 30% of total area being worked at the time of Stevens to 15% of area in 1988. In 1988 two mineral types accounted for over half of all workings with no provisions for restoration, "other minerals" and limestone.

The figure for "other minerals" refers largely to peat workings within Doncaster MBC and Somerset CC.

Table 1.1 :   Area of surface mineral workings with no provisions for restoration and aftercare

Mineral	% of area having no restoration conditions Stevens (1976)	% of area having no restoration provisions DoE (1982)	% of area having no restoration provisions DoE (1988)
	%	%	%
Chalk	26	18	39
China Clay	94	66	62
Clay/Shale	65	9	8
Coal (opencast)	-	<1	<1
Gypsum/Anhydrite	51	15	14
Igneous Rock	48	30	18
Ironstone	-	<1	<1
Limestone	58	38	35
Oil/Gas exploration	-	-	4
Oil/Gas production	-	-	3
Sand and Gravel (construction)	16	5	5
Sand industrial	-	10	7
Sandstone	27	30	31
Slate	97	97	84
Vein Minerals	-	12	42
Others	-	47	66
Overall	30	15	15

Note : (1) Calculated from Tables 12.2 and 7.1 of DoE 1988 Survey for Land for Mineral Workings in England.

Stevens found that of those sites having satisfactory conditions, 23.4% had been worked and conditions complied with, 10.7% had been worked and the conditions not complied with and 65.9% were still working or not yet started. By mineral type Stevens evidence suggested that clay/shale, coal (excluding NCB opencast), igneous rock, limestone and other minerals had experienced default of conditions in excess of 50% of sites completing operations.

Stevens' detailed examination of reasons for failure shown in Table 1.2. Stevens found that financial failure accounted for 3.4% of failure cases. Other reasons were wide-ranging, but the largest single reason was the failure to either monitor or enforce restoration conditions. Stevens concluded that procedural difficulties and delay involved in implementing the enforcement system were sufficient to deter many planning authorities from taking action. Other failures of the planning system including loosely worded, incompatible and unrealistic restoration conditions.

Stevens also identified a range of diverse technical problems leading to failure such as shortage of fill.

Table 1.2 : Causes of major failure to observe satisfactory restoration conditions

Major cause of failure	Number of sites	% of cases
Loosely worded conditions prevented full compliance	7	4.8
Enforcement in hand	6	4.1
Working now suspended; taken over by major company	10	6.9
Operator claims he has observed conditions	2	1.4
Still being reworked/possibility of reworking	24.5	16.9
Liquidation/bankruptcy of operator	5	3.4
Site not yet worked and/or dereliction from earlier workings	14	9.7
Working inadequacy monitored and restoration conditions not enforced	42	29.0
Delayed by shortage of fill but likely to be restored in due course	11	7.6
Original restoration conditions incompatible with present or envisaged usage	11	7.6
Restoration not carried out by agreement with LPA in anticipation of further development which did not materialise	2	1.4
Derelict due to circumstances arising after adequate restoration	6	4.1
Unrealistic conditions	1	0.7
Information given insufficient to identify causes of non-compliance	3.5	2.4
Total	145	100

Source : Stevens Report 1976

County Planning Officers' Society

The County Planning Officers' Society (CPOS) examined minerals sites ceasing operation and their restoration performance in the period 1974-82. Survey findings were based upon responses received from 48 mineral planning authorities in England and Wales. A summary is provided in Table 1.3.

In the opinion of the MPAs 60.6% of sites had been or were expected to be restored satisfactorily where conditions would not present a problem for present and future uses.

Table 1.3 :   Restoration performance 1974-1982

Category of restoration	Number of sites	% of total site
Satisfactory restoration	599	60.6
Unsatisfactory restoration	89	9.0
New development	106	10.7
Operations suspended	107	10.8
No restoration	88	8.9
Total number of sites	989	100

Source : County Planning Officers' Society (1985)

The CPOS found that reasons for failure to achieve satisfactory restoration were complex although by far the most frequently mentioned factors, accounting for 42% of failure cases, were either the absence or inadequacy of planning conditions relating to restoration. Inadequate monitoring and supervision of sites was mentioned in a further 9% of unsatisfactorily sites.

The next most frequent reason, accounting for 23% of failure cases, related to site specific technical difficulties.

The CPOs study found that financial failure had been a principal reason in 9.3% of cases of failure.

Department of the Environment 1982 and 1988 Surveys

As indicated in Table 1.1, the DoE undertakes periodic surveys of land for mineral workings with information provided by mineral planning authorities. In addition to providing general statistics on the extent of mineral workings, the 1988 Survey gives an indication of the planning status of mineral workings with respect to restoration and aftercare conditions.

In 1988 there were 15850 ha of surface mineral workings with no restoration conditions nor other provisions for restoration, equivalent to 16% of the total working area. Nominally this figure shows little change since 1982 although this may reflect an increase in the standards of environmental safeguards and performance expected by MPAs.

Summary of Previous Evidence

As regards the status of mineral planning permissions in relation to conditions, the principal conclusion has been:

- a significant minority of sites have no or unsatisfactory restoration conditions, but the proportions of sites in this category is falling.

As regards restoration performance, the principal conclusions have been:

- a significant incidence of problems derive from failure to implement the planning system, including inadequate restoration conditions, lack of monitoring and failure to take enforcement action;

- there has been a low incidence of failure to restore due to financial failure;

- technical reasons for failure to restore are associated with a wide variety of site-specific difficulties.

1.4 An Outline of Existing Planning Procedures for Restoration Following Mineral Workings

Mineral workings like other forms of development fall within the scope of development control under the planning acts. Unlike most other developments which improve land or increase its value, mineral development is a destructive process. The effects on land and the environment need to be repaired and consideration needs to be given to appropriate afteruse. With few exceptions, neither the operator nor the landowner has an economic incentive in the site to undertake satisfactory restoration and aftercare after extraction and reliance has to be placed on the planning system.

For disused mineral sites which have not been restored, including those which pre-date the 1947 Town and Country Planning Act, reclamation is likely to occur only with public sector financial assistance. The Derelict Land Grant is Government's principal policy instrument for this purpose in England.

The long time span of many mineral permissions means that planning and environmental standards may change during the life of a permission and contemporary standards usually exceed those relating to conditions imposed on many older permissions. This problem has been particularly acute because of non-existent or rudimentary and inadequate standards of restoration imposed in conditions of older permissions which are currently coming to an end. It is advantageous, therefore, to consider current practice and recent legislation before reviewing the nature of any outstanding limitations.

Restoration and Aftercare Provisions under the 1981 Minerals Act

The minerals planning system was updated by the Town and Country Planning (Minerals) Act, 1981, which had the effect of amending the Town and Country Planning Act, 1971. A number of new powers and duties were conferred which were directed towards the environmental problems of mineral extraction, and in particular, to securing the restoration and aftercare of sites after extraction has been completed. These included:

- powers for planning authorities to impose aftercare conditions with a five year obligation, after land has been restored, to bring the land up to the necessary standard for agricultural, forestry or amenity use;

- a provision that all mineral permissions should be time-limited; with those existing as at 1982 timed to expire after 60 years, by 2042, unless otherwise specified;

- the creation of specialist mineral planning authorities in England and Wales.

The 1981 Minerals Act also introduced provisions to tackle the problems of inadequate restoration conditions on older permissions. This took the form of a general duty for planning authorities to review all mineral permissions in their areas, although the timing of these reviews was left to the authorities' discretion. For sites identified in the review requiring updating of conditions, the Act introduced extended powers of revocation and discontinuance, and for sites which had not been worked for a number of years new powers of prohibition and suspension.

Planning authorities were therefore given the powers to seek higher environmental standards in respect of development already underway or having permission, and to bring past permissions into line with modern requirements.

The cessation of mineral extraction before the mineral resources have been fully exhausted and the consequent failure to undertake restoration has posed a major specific problem for planning authorities. The Act empowered authorities to require that works be carried out upon the cessation of extraction, and that these should be works of amenity in the case of temporary cessation and full restoration and aftercare in the case of permanent cessation.

New compensation provisions were made for the exercise of these powers. These were set such that the operator was required to bear a proportion of the cost of works required for protecting the environment thereby reducing, but not normally eliminating the cost burden falling on authorities. Implementation of the Act has not carried with it any additional resources to planning authorities for the implementation of its objectives.

Interim Development Orders and the Planning and Compensation Act 1991

Interim Development Orders (IDOs) were permissions granted after 21 July 1943 and before 1 July 1948 which have been preserved by successive Planning Acts as valid permissions in respect of development which had not been carried out by 1 July 1948. More often than not these have posed problems for planning authorities due to an absence of or inadequate restoration conditions. Additionally, they were not required to be included on the general register of mineral planning permissions.

The Planning and Compensation Act, 1991 required holders of such permissions to apply to the mineral planning authority for registration of the permissions by 25 March, 1992 and subsequently to apply for determination of working and restoration conditions which will be imposed on the registered permission without compensation.

Government advice in MPG9 on the setting of conditions on working suggests that a distinction should be made between "dormant" permissions, where full modern conditions will be appropriate, and "active" permissions where conditions should deal with the environmental and amenity aspects of working the site without affecting the economic structure of the operation.

Enforcement

Even where restoration conditions of a planning permission are satisfactory by contemporary standards restoration may fail to take place or may fail to meet in full the conditions imposed as a result of operator default. In this event planning enforcement procedures come into effect, and the planning authority has at its disposal a number of instruments which are designed to achieve compliance with the conditions imposed.

Breach of restoration or aftercare conditions on a minerals site is not in itself an offence. Legislation provides for a buffer between breach and prosecution in the form of enforcement action. Action is exercised by the issue of a notice by the planning authority specifying the matters alleged to constitute a breach, the steps required to remedy the breach, the date on which the notice should take effect, the period in which any steps should be taken and any other additional regulatory matters. Its purpose is remedial rather than punitive, and a power of prosecution exists only in respect of failure to observe the requirements of the notice. Planning authorities may also take almost immediate action by serving a stop notice in addition to an enforcement notice. A stop notice might be used to prevent further adverse action by the operator which would be irreversible. Compensation may be payable if the notice is subsequently quashed (on grounds other than the grant of permission by the Secretary of State).

As Stevens concluded, enforcement powers have not always achieved their desired result. Planning authorities have sometimes encountered difficulties in pursuit of operators in cases of financial failure and other instances of disappearance, or their successors in title. And technical difficulties, disputes over conditionality and the time required and costs of monitoring and enforcement have caused problems and have resulted in restoration failure.

Following the recommendations of the Carnwath Report, **Enforcing Planning Control** (HMSO, 1989), a number of reforms to enforcement action have been implemented in The Planning and Compensation Act, 1991. These new provisions include:

- a substantial increase in the level of summary fines for failure to comply with the requirements of an enforcement notice or the prohibition in a stop notice;

- a discretionary procedure for planning authorities to obtain information about activities where a breach of control is suspected in the form of a 'planning contravention notice';

- a summary procedure for enforcement of planning conditions known as a 'breach of condition notice'.

Current Financial Responsibilities for Restoration

The Planning Acts do not refer to expenditure on restoration, and aftercare, but the 1981 Mineral Act places the onus of financial responsibility firmly on the person who last had the benefit of winning and working the minerals to restore sites and hence to ensure that adequate provision is made for this purpose. If default occurs, the Planning system has remedies but the incidence of financial liabilities remains with the operator, or successor in title.

Current Planning practice and policy advice in relation to planning conditions prohibits local authorities from making financial contributions a condition of planning permissions.

Some local authorities have adopted policies to require operators to make financial guarantees. However, to give practical effect to these policies local authorities can only enter into voluntary Agreements with operators. To date, only a few Agreements have been made at the time of granting mineral planning permissions although the practice is growing.

2 CURRENT PRACTICE AND EXPERIENCE REGARDING MINERALS RESTORATION

2.1 Surveys of Local Authorities and Mineral Operators

Postal surveys of mineral planning authorities in England and Wales and planning authorities responsible for minerals in Scotland, and a large sample of mineral operators were undertaken in the second half of 1991.

The survey of local authorities was intended to provide statistical information in relation to the restoration of sites in the period 1982 to 1990 ie. in the period since the implementation of the 1981 Act. The information sought in relation to these sites included:

- whether restoration is happening to a good and acceptable standard and whether there are significant differences related to different sectors of the industry or different areas of the country;

- whether there are any problems with the speed with which restoration is occurring;

- the scale of default by the minerals industry on restoration of sites and the reasons.

The survey was also designed to facilitate comparisons with the two previous surveys by Stevens (1976) covering the period 1943-1972 and the survey by the County Planning Officers' Society CPOS (1985) covering the period 1974-1982.

In addition, the postal survey provided some clarification of English authorities' approaches to the 1982 and 1988 Department of the Environment Surveys of Land for Mineral Workings, particularly in relation to the categorisation of sites as having "unsatisfactory restoration conditions" in the DoE 1988 Survey. Supplementary information was also sought from MPAs on opinions and judgements on certain aspects of minerals planning, restoration and aftercare. The questionnaire used is at Appendix 1; a full list of respondents is provided at Appendix 2. The response rate for the planning authority postal questionnaire at 58% overall was high for a study of this nature, bearing in mind the quantity of information requested. Details are provided in Table 2.1.

267 mineral operators were sent postal questionnaires including a representative selection of national and local operators of different mineral types and geographical distribution. The survey of mineral operators explored opinions and judgements about the working of the minerals planning system in relation to restoration and aftercare matters and operators' experiences. A copy of the mineral operators' questionnaire is at Appendix 1; a full list of respondents is provided at Appendix 2. The response rate to the operators' postal survey was 35% (93 responses). The survey also achieved a good regional coverage by both national and local operators and by mineral type.

The postal questionnaire survey was supplemented by 20 follow-up personal interviews with planning authorities and 20 interviews with mineral operators, a full list of which is provided at Appendix 3. The interviews were intended to provide further details of restoration performance including: clarification of the results of the postal surveys; details of development control experience including the setting of conditions, monitoring and enforcement; implementation of the provisions of the 1981 Minerals Act, and the use of planning Agreements.

Table 2.1 : Response rates for the local authority postal survey

Country	Administrative level	Total number of relevant planning authorities	Completed responses (percent of total)
England and Wales	Counties	44	29 (66%)
	Metropolitan[1]	34	20 (59%)
	National Parks	3	3 (100%)
Scotland	Regional Councils	6	2 (33.3%)
	Districts	41	20 (49%)
Total		128	74 (58%)

Notes: 1   excludes Liverpool and Wirral as in 1988 Minerals Survey
            includes London Boroughs of Bexley and Sutton, but no other London Boroughs

Restoration Performance : Evidence from the Local Authority Surveys

In the postal questionnaire local authorities were asked to judge the level of restoration of mineral sites ceasing in the period 1982-1990 according to the categories "satisfactory restoration", "unsatisfactory restoration" or "no restoration" by mineral type. The results of this analysis is presented in Table 2.2.

In answering this question authorities were advised through guidance notes to take account of what would be reasonably achievable having regard to the nature of planning conditions, the condition of the site prior to mineral working, the physical constraints of the site and whether or not the condition of the site is suitable for the intended afteruse. No restoration refers to sites or part sites where mineral working has ceased or operations completed and no attempt has been made to restore the site, including sites where there were no obligations for restoration.

The standard of satisfactory restoration varied from authority to authority. However, authorities tended to put primary emphasis on visual appearance and sustainability of afteruse. Other factors considered include the extent to which planning conditions have been fulfilled, the productivity of afteruses (principally relating to agriculture), ecological interest, quality and sustainability of vegetation, and acceptability to local communities.

Table 2.2 : Restoration performance 1982-1990

Mineral Type	Total number of sites	Sites satisfactorily restored		Sites unsatisfactorily restored		Sites not restored	
		Number	%	Number	%	Number	%
Chalk	46	26	56.5	1	2.2	19	41.3
China clay	0	0	0.0	0	0.0	0	0.0
Clay/shale	84	65	77.4	10	11.9	9	10.7
Coal (deep mined)	165	61	37.0	48	29.1	56	33.9
Coal (opencast)	220	187	85.0	26	11.8	7	3.2
Gypsum/Anhydrite	1	1	100.0	0	0.0	0	0.0
Igneous Rock	16	6	37.5	4	25.0	6	37.5
Ironstone	2	2	100.0	0	0.0	0	0.0
Limestone	72	43	59.7	14	19.4	15	20.8
Oil/Gas exploration	86	84	97.7	0	0.0	2	2.3
Oil/Gas production	3	3	100.0	0	0.0	0	0.0
Sand and Gravel (construction)	539	446	82.7	49	9.1	44	8.2
Sand (industrial)	57	41	71.9	12	21.1	4	7.0
Sandstone	28	12	42.9	12	42.9	4	14.3
Slate	3	0	0.0	0	0.0	3	100.0
Vein minerals	95	56	58.9	18	18.9	21	20.1
Others	47	31	66.0	8	17.0	8	17.0
Overall	1464	1064	73.0	202	13.9	198	13.6

Note : Numbers may not sum due to rounding

Approaches to restoration vary considerably between authorities. A number of authorities apply standard conditions or are reactive to operator proposals. Others take a more proactive approach; one authority was, for instance, examining the final face configuration of quarries, while others had consciously sought to introduce new ideas and approaches to some of the smaller, less competent operators.

Table 2.2 shows that in the opinion of the local authorities overall 73% of mineral sites were satisfactorily restored over the period 1982-1990 although this varies considerably by mineral type.

Opencast coal and sand and gravel sites for construction have the highest rates of satisfactory restoration with in excess of 80% of sites being restored to a satisfactory standard. No Opencast Executive sites were left unrestored and in the few cases that were classed as unsatisfactory, issues of concern which were raised related to the final afteruse rather than the technical standard of restoration.

In the view of the planning authorities the highest rate of unsatisfactory or no restoration occurs on deep mined coal sites, although this record is closely followed by igneous rock sites. In addition, a significant proportion of clay/shale, other rock sites (limestone and sandstone) and vein minerals have not been restored to the planning authorities' satisfaction.

The picture changes somewhat when considering the scale and thereby the overall impact of restoration failure. While the rate of failure on sand and gravel sites is low, these sites account for 24% of the total number of sites unsatisfactorily restored and 22% of cases of no restoration.

Authorities suggested that operators' technical competence, creativity and standard of restoration had improved in recent years and operators often proposed imaginative restoration schemes. Concern was expressed however, about the restoration of existing hard rock, (particularly limestone) and metalliferous mineral sites. In many cases, the final restoration proposals were unclear and there was doubt that operators have made adequate financial provision.

Authorities were asked to classify restoration performance by type of operator and the results of this analysis are presented in Table 2.3. This reveals no significant difference between national and local operators. National operators were defined for common minerals as having operations in several regions or a large market share for specialist minerals (e.g. china clay or fluorspar) or national importance.

Table 2.3 : Restoration performance by type of operator

Type of operator	Total number of sites	% satisfactory restoration and/ or aftercare by operator	% unsatisfactory restoration and/ or aftercare by operator	% no restoration by operator
National Operator	734	74.3	11.3	14.4
Local Operator	520	69.8	17.5	12.7
Total	1254	72.2	14.1	13.6

Note : Total differ from those in Table 2.2 and elsewhere due to incomplete responses

Authorities indicated in the follow-up interview surveys with a perception that there were invariably less responsible operators that failed to perform well. Whilst they were often small this did not reflect the performance of most of the small or medium sized operators. Nevertheless, authorities are dependent upon the technical competence of the operators which has tended to be concentrated within the medium and large operating companies.

Table 2.4 provides a breakdown of reasons given for unsatisfactory or no restoration.

Table 2.4 shows that the failure to implement the planning system has been a significant reason for failure to restore. This includes the absence of conditions or other provisions for restoration, unrealistic planning conditions, loosely worded planning conditions and lack of monitoring which together account for 34.6% of failure cases.

The incidence of financial failure is low at 5.0%.

Table 2.4 :   Reasons for unsatisfactory or no restoration and/or aftercare

Reason	No of sites	% of sites with unsatisfactory or no restoration
(a) unrealistic planning conditions	19	3.9
(b) loosely worded planning conditions	33	6.8
(c) no requirement to restore	80	16.5
(d) lack of monitoring	36	7.4
(e) financial failure of the operator	24	5.0
(f) suspension of activities	74	15.3
(g) shortage of fill materials	18	3.7
(h) possibility of reworking the site	47	9.7
(i) inadequate practices by mineral operators eg. soils damaged or lacking	66	13.6
(j) slowness or delays not otherwise included	87	18.0
Total	484	100.0

Note :   More than one reason may apply to sites

Other reasons for failure include a range of technical or site specific problems including inadequate practices by mineral operators, operator negligence and lack of expertise. They also include circumstances beyond an operator's control. This is clearly an area of judgement but includes problems of a wide range of stated reasons such as a profuse growth of weeds, contamination by infill, uneven settlement or poor drainage. Circumstances contributing to slowness or delays within category (j) include changes in ownership; technical problems such as contaminated fill materials; gas and subsidence; changes in operator personnel and, in the case of peat, fragmented patterns of land ownership.

Reflecting a finding of the DoE 1988 Survey of Land for Mineral Workings, authorities indicated that a significant proportion of sites had "unsatisfactory restoration conditions". Particular problems arise because of differences between authorities intentions when setting conditions and their poor and ambiguous wording. While planning conditions are often suited to achieving negative purposes (such as restricting hours of working) there are more difficulties when used to specify positive steps that an operator must take to achieve planning objectives and invariably need to be expressed in great detail.

Box 1 : Example restoration conditions taken from a recent permission for a quarry

CONDITIONS

1. So much of the development the subject of this permission as consists of the winning and working of rock must cease not later than 10th May, 2007.

2. At least one year before the completion of quarry workings at level number X (Drawing XXXXXXX), a fully detailed scheme shall be submitted for the approval of the planning authority showing the proposed means of collection, treatment and disposal of contaminated surface water arising from within the quarry during all phases of quarry working below level number X. No working beyond level number X shall be permitted until such scheme has been approved in full by the planning authority.

3. Top soil and sub soil overburden stripped from any part of the site shall be tipped and stored in locations to be previously approved in writing by the planning authority. All such material shall be conserved for use in site landscaping and restoration works. No other material removed from any part of the site, including material produced during the construction phase, shall be tipped on the site or adjoining areas without the prior written approval of the planning authority. Such material includes weathered rock, dust, fine particulate or other forms of spoil produced as a result of the winning and working of the rock.

4. All plant and machinery which is no longer required for the winning and working of rock or for any other purpose shall be stored in a manner and location to be agreed in writing by the planning authority in readiness for its removal from the site such removal to be effective within three months of such redundancy or within such longer period as the planning authority may agree.

5. The developer shall from time to time as required by the planning authority submit for the approval of the planning authority and thereafter implement a detailed programme of progressive and continuous reinstatement of the site, to the intent that all land disturbed by construction works shall be reinstated to the satisfaction of the planning authority within six months of completion of that element of construction which occasioned such disturbance and all temporary buildings, plant, machinery and other temporary works shall be removed from the site and form adjoining land within three months of them ceasing to be used.

6. In conjunction with the reinstatement programme referred to in condition X there shall also be submitted for the approval of the planning authority and thereafter reviewed as required by them a scheme for the final restoration of the site which shall be implemented by the developer within six months of the cessation of the winning and working of minerals and, without limiting the generality of the foregoing, such scheme shall require the developer to carry out the following works to the satisfaction of the planning authority:-

   (a) the entire removal from the site and from adjoining land of all buildings;
   (b) the reinstatement of all land disturbed by and exposed by such removal as is described in (a) above, such reinstatement to include the tipping and contouring of soil, grass seeding and tree planting;
   (c) the reinstatement of all land within and adjoining the site which has been disturbed otherwise than by the removal of buildings, plant, machinery, structures and erections and which the planning authority considers requires restoration by the spreading of soil, infilling, contouring and planting with grass and/or trees;

7. Also in conjunction with the reinstatement programme referred to in condition XX, there shall be submitted for the approval of the planning authority a management plan indicating the manner in which all areas of tree planting, stockproof fencing, access and surface water drainage shall continue to be cared for and maintained following cessation of mineral working and the developer shall thereafter maintain the site in accordance with such approved management plan.

> 8. In the event of construction works or the winning and working of rock on the whole or in any part of the site being abandoned at any time, such further work shall be carried out as the planning authority may require in order to restore and reinstate the site in a manner which shall minimise the environmental impact of such abandonment. Abandonment shall be deemed to have occurred if work has ceased to be carried on in such active and continuous manner as indicates to the planning authority an intention, by reason of such cessation, to abandon. Notwithstanding the foregoing, where the whole development is placed on a care and maintenance basis sufficient to prevent deterioration in visual appearance and safety the planning authority may defer their requirement for restoration and reinstatement of the site until the expiry of a period of six months from the date of any such cessation or until of such longer period as the planning authority shall approve. Without prejudice to the foregoing, the planning authority may nevertheless require reinstatement and restoration of the site in accordance with conditions 5, 6 and 7 of this permission.

Problems with many current conditions include imprecision such as, "Restoration of the site will commence when working is substantially complete". A second problem is of impracticality such as, "The operator will respread topsoils throughout the site to a depth of 1000 mm" (when original pre-working depths were 500 mm). A third problem is of inflexibility, mostly occurring where the original afteruse is no longer considered appropriate for the particular site.

Box 1 provides a typical example of current practice regarding the setting of conditions. It illustrates the difficulties of setting precise conditions on long term sites. For comparison, Box 2 provides an equivalent example for a short-term operation where conditions require the early submission of a detailed plan of working and restoration.

In the opinion of many planning authorities, best practice is to follow the advice offered in MPG7 and to undertake careful site investigations, and undertake discussions of working and reclamation proposals prior to the formal submission of an application. Of course, many sites pre-date MPG7 in the dates of their permissions. Planning conditions will also be closely referenced to detailed restoration plans, drawings and material budgets.

Prominent amongst past problems are failures on the part of operators to make adequate site investigations prior to working and of planning authorities to clarify their intentions prior to setting conditions. Greater use might also be made of reserved matters in the final restoration scheme, although in these cases authorities indicated that it is still important to establish a comprehensive and well planned scheme of working the site at the outset in order to maximise the options for final restoration.

Monitoring Activities by Local Authorities

Within authorities monitoring is generally the responsibility of specialist mineral officers who in England are responsible typically for both minerals and waste disposal. Staff numbers devoted to development control in authorities vary from between one and six officers working full time on minerals. There is no clear relationship between numbers of officers and caseload. A large number of officers indicated that their caseload has increased in recent years. Many of the metropolitan districts, particularly in the major minerals extracting areas, have developed specialist mineral groups either individually or in some cases on a collective basis.

In Scotland, responsibility varies between districts and regional councils and monitoring is usually undertaken as part of other development control functions. Activity and interest in minerals planning and the need for high standards of restoration varies considerably and this is reflected in the variability of effort devoted to monitoring. For example, a larger proportion of Scottish authorities than in England and Wales have sought bonds or other security for restoration by Agreement but nevertheless, in Scotland there are many authorities that have been less concerned to take action, reflecting low populations, the historical acceptance of quarrying and local economies dependent on extraction activities.

Specialist monitoring and enforcement staff tend to come from diverse backgrounds and include, in particular, former policemen and mining engineers.

The regularity with which authorities monitor minerals sites varies considerably. A number of authorities in the survey undertake regular monitoring of minerals sites involving routine site visits and systems for prioritising visits ranging from a simple diary to sophisticated computer systems. For others monitoring is undertaken on an ad hoc basis often concentrating on particular kinds of sites including those where the operator's ability to restore was in doubt or at crucial points in the restoration process.

26.5% of MPAs that responded in the survey in England and Wales and a similar figure for Scotland produce regular monitoring reports on the condition of mineral sites within their areas. Reasons stated for not producing such reports include lack of staff resources (52.9% England and Wales) and lack of necessary expertise (8.8% England and Wales).

Inspection of monitoring reports suggests that the primary emphasis of site visits is on the visual appearance of the site and compliance with visually sensitive conditions. Comments made during interviews suggest that monitoring is regarded as necessary although many authorities lack the staff resources to undertake it properly. A number of respondents suggested that the focus of current monitoring activity is on known problem sites. A greater degree of monitoring is required on sites at critical periods. As one respondent suggested:

> "I doubt whether most Planning Authorities have trained staff with the time to inspect and attend meetings on a mineral working site at a frequency of more than once a week for a sustained period. Unfortunately there are critical periods, such as when soils are being stripped, treated and respread, when the quality of restoration can be adversely and perhaps irrevocably reduced in a matter of a day or two".

Enforcement Action by Local Authorities

Planning authorities' approach to enforcement closely follows Government circulars and policy advice. Enforcement action is taken as "a last resort" and most regard the establishment of good relationships with operators as the most important factor in securing compliance with conditions. Few authorities have used enforcement procedures extensively even when faced with major restoration problems.

Box 2 : Example restoration conditions for short term operations on an opencast coal site

1. No coal shall be extracted from the site after 30 April 1995 and the date of cessation of extraction operations shall be notified to the Mineral Planning Authority in writing upon or immediately following cessation. Thereafter the site shall be restored in accordance with the provisions of condition (11) to this permission.

2. The restoration of the site in accordance with the provisions of condition (11) to this permission shall be completed as soon as is practicably possible, having regard to weather and ground conditions, and within six months from the site shall be completely restored no later than 29 June 1995 except as may otherwise by agreed in writing by the Mineral Planning Authority. The applicants programmed date for completion of the restoration shall be notified to the Mineral Planning Authority in writing no later than six weeks prior to that programmed date.

3. Prior to the commencement of soil stripping operations in compliance with conditions (7) and (8) to this permission the access laid out in accordance with the specifications and design shown on the applicants Drawing No: XXXXX and thereafter maintained in this condition until required to be removed in compliance with condition (10) to this permission.

4. Prior to the commencement of soil stripping operations in compliance with conditions (7) and (8) to this permission the site boundary hedge adjoining High Lane shall be cut back to the root line on the highway frontage.

5. Prior to the commencement of soil stripping operations in compliance with conditions (7) and (8) to this permission a plan showing details of the proposed location, extent, height, form and treatment of all mounds of topsoil and subsoil and overburden dumps not the subject of previous approval shall be submitted to the Mineral Planning Authority. Thereafter the mounds shall be constructed and treated in accordance with the details as approved or modified or as may subsequently be agreed in writing by the Mineral Planning Authority.

6. Prior the commencement of soil stripping operations in compliance with conditions (18) and (19) to this permission a plan to a scale of not less than 1:500 showing details of the location, extent, height, form, materials of construction, treatment to and maintenance of the proposed screen mound indicated on the applicants Drawing No. 2 shall be submitted to the Mineral Planning Authority. Thereafter the mound shall be constructed, treated and maintained in accordance with the details as approved or modified or as may subsequently be agreed or required in writing by the Mineral Planning Authority.

7. Prior to the excavation of the overburden all available topsoil shall be separately stripped from the area to be excavated, from the proposed sites of subsoil mounds and overburden dumps and from all areas likely to be traversed by heavy plant and machinery. All topsoils stripped from the site shall be separately stocked in accordance with such details as may be approved under the provisions of conditions (5) and (6) to this permission until required for the restoration of the site in compliance with condition (11) to this permission.

8. Prior to the excavation of the overburden all available subsoil shall be separately stripped from the area to be excavated. All subsoil stripped from the site shall be separately stocked in accordance with such details as may be approved under the provisions of conditions (5) and (6) to this permission until required for the restoration of the site in compliance with condition (11) to this permission.

9. As far as is practicable the stripping and movement of soils shall only take place during periods of fine weather when the soils are in a suitable dry and friable condition.

10. At such time as they are no longer needed in connection with the opencasting and restoration of the site, the access shall be removed, the access mouth shall be physically stopped up and the affected area within highway limits shall be reinstated to such design and construction as may be specified by the Mineral Planning Authority in consultation with the Local Highway Authority.

11. Within two months of the date of this permission or within such other period as may be agreed a detailed scheme in respect of the restoration of the whole site shall be submitted to the Mineral Planning Authority. The scheme shall provide for the restoration of the site to a suitable condition to facilitate the approved use of the existing opencast site areas for fishing ponds and amenity purposes and shall amongst other matters (except as may otherwise be agreed in writing by the Mineral Planning Authority), include details of the following:
   (a) the replacement of the overburden within the excavations;
   (b) a detailed restoration contour plan;
   (c) the quantity of available subsoil on the site and the quantity of subsoil to be imported;
   (d) the quantity of available topsoil on the site and the quantity of topsoil to be imported;
   (e) the spreading of the soils in their correct sequence including details of the proposed final location and depths of subsoil and treatment thereto;
   (f) adequate drainage facilities;
   (g) the formation of a new access to the site in accordance with the requirements of condition (4) of the planning permission dated 21 January 1981 relating to application XXXX,
   (h) adequate provision of suitable surfaced space for on site parking, turning and manoeuvring facilities;
   (i) adequate provision for vehicular access for maintenance purposes;
   (j) adequate provision for suitable surfaced on site pedestrian accessways.

Thereafter the scheme shall take place, within the time limit imposed in condition (3) to this permission, in accordance with the details as approved or modified or as may subsequently be agreed in writing by the Mineral Planning Authority in consultation with XXXX Borough Council.

12. There shall be no importation of fill materials in connection with the restoration of the site other than the importation of soils in accordance with the scheme required under the provisions of condition (11) to this permission.

13. Within two months of the date of this permission or within such other period as may be agreed a detailed scheme in respect of the landscaping of the whole site shall be submitted to the Mineral Planning Authority. The scheme shall provide for the landscaping of the site until and without prejudice to restoration to a condition suitable to facilitate the approved use of the existing opencast site areas for fishing ponds and amenity purposes and the use of the former sewage works area for amenity or recreational purposes and shall, amongst other matters, include details of the following:
   (a) the seeding to grass of appropriate areas in the first available seeding season;
   (b) the planting of hedgerow plants, in the first available planting season, to make good the boundary hedge along the frontage of the site with Back Lane;
   (c) the planting of trees and shrubs, including details of location, numbers, planted size and species, in the first available planting season;
   (d) the provision of fencing and gates, including details of the location and types of fencing and gates to be erected and protective treatment thereto;
   (e) a programme of implementation.

Thereafter the scheme shall be implemented in accordance with the details as approved or modified or as may subsequently be agreed in writing by the Mineral Planning Authority in consultation with the Borough Council.

14. The first five years following implementation, all planting shall be maintained in accordance with the principles of good forestry and husbandry and any hedgerow plants, trees or shrubs which die or become seriously damaged or diseased shall be replaced.

15. An aftercare scheme, providing for such steps as may be necessary to bring the site to the required standard for use for amenity (ie. when the land is suitable for sustaining tress, shrubs or plants), shall be submitted for the approval of the Mineral Planning Authority not later than six weeks prior to the applicant's programmed date on which it is expected that the restoration of the site required by condition (11) to this permission will be completed. The submitted scheme shall specify the steps to be taken and the periods during which they are to be taken.

16. Thereafter the aftercare of the site shall be carried out in accordance with the aftercare scheme as approved or modified or as may subsequently be agreed in writing by the Mineral Planning Authority in consultation with the Borough Council.

Comments on implementing enforcement action was broadly similar to those reported by Carnwath. In particular, comments focus upon difficulties arising from:

- complexity and delay;
- lack of urgency;
- inflexibility.

Complexity and delay arise from the problems of establishing that a breach of conditions has occurred. Planning conditions have often been imprecise or open to interpretation, which in turn allows the alleged contravener to put off effective action through the appeals procedure. This results in a perceived lack of urgency within the system as a whole.

Very few authorities have used the stop notice procedure to assist with securing satisfactory restoration. Authorities thought the scope of this instrument was limited, and are concerned about the risk of compensation payments.

In most authorities, enforcement action is handled by a separate legal department. Planning officers often lose sight of the process once initiated. They would prefer to take action within their own department.

Often enforcement action is not taken because of the cost implications for the authority, particularly where the breach is limited in its public or visual impact.

Where a breach has occurred due to operators' financial failure the obligations to restore where planning conditions exist passes to the successor in title, typically the landowner. In the interviews with authorities it was generally regarded as unlikely that landowners could be pursued successfully with enforcement action and this was not generally undertaken.

Many authorities that have used enforcement action successfully could identify no methods of improving the mechanism itself and suggested that the knowledge that the system can and will be used is often sufficient to discourage default.

Table 2.5 shows the extent to which enforcement action has been used or is proposed to be used by authorities responding to the postal questionnaires to achieve satisfactory restoration. This reveals that enforcement action has only been taken on 35 of the survey sites between 1982 and 1990, equivalent to only 8.8% of all failure cases. It was thought by local authorities to have been possible to undertake enforcement action only in a further 13.3% of cases.

Implementation of the 1981 Minerals Act by Local Authorities

73.0% of responding authorities in England, Wales and Scotland considered that the powers contained in the 1981 Minerals Act had improved the likelihood of satisfactory restoration. 39.5% of authorities had undertaken the Review covering a total of 1096 sites or permissions of which 314 had been identified as having scope for adding or modifying restoration conditions. However, actions have been taken in only 19 cases.

Table 2.5 : Level of enforcement action in relation to sites identified as having unsatisfactory or no restoration

Enforcement Status	Number of sites	% of total sites
Enforcement action taken, or proposed to be taken	35	8.8
Thought possible for the planning authority to institute enforcement action	53	13.3
Sites where enforcement action not considered	312	78.5
Total	400	100.0

Almost two-thirds of authorities suggested that there were constraints reducing the effectiveness of the provisions in the 1981 Minerals Act, with 42% identifying the major constraint as a general one of lack of financial resources for implementation. A small number of authorities cited the problems in achieving restoration in cases where working had not taken place for a number of years and where a prohibition order had been served, and where the site had no or only limited restoration conditions.

Comments made by authorities indicate that the result of the financial constraint is that the powers to update permissions are used rarely, or only in respect of comparatively minor works.

Many authorities have found it easier to use planning applications for the extension of existing sites or for new sites to modernise the permissions on older sites by introducing new or updated restoration or aftercare conditions, without compensation, through the use of planning Agreements under Section 106 of the Town and Country Planning Act 1990 (formerly S52 of the 1971 Act) and S.33 of the Local Government (Miscellaneous Provisions Act) 1982 to supplement restoration and aftercare conditions. Operators have also frequently been encouraged to seek consolidating permissions.

Authorities interviewed suggested that the number of sites where provisions for restoration have been updated using these mechanisms is substantially greater than those updated using 1981 Minerals Act powers.

Most new permissions included aftercare conditions. However, a small number of authorities expressed doubt as to the practical realities of being able to enforce aftercare conditions. In some cases title would revert back to the original landowner or passed to an alternative successor when restoration works were complete.

Statistical Evidence from the Mineral Operator Survey

The postal survey of mineral operators achieved a wide geographical and minerals coverage of sites. Table 2.6 provides a summary of survey sites by operating regions broken down by type of operator, while Table 2.7 provides a summary of sites ceasing operations in the period 1982-1990.

Table 2.6 : Regional distribution by type of operator

	National Operators		Local Operators	
	Number	% of total	Number	% of total
South West	17	13.4	11	20.4
South East	16	12.6	13	24.1
East Anglia	8	6.3	3	5.6
East Midlands	17	13.4	4	7.4
West Midlands	12	9.4	7	13.0
North West	14	11.0	2	3.7
Yorkshire and Humberside	12	9.4	3	5.6
North	12	9.4	1	1.9
Scotland	10	7.9	10	18.5
Wales	9	7.1	0	0.0
Total	127[1]	100.0	54	100.0

Note : 1 Refers to regions where national operator has a presence, hence total number exceeds the number of operators.

Table 2.7 : Number of operator sites ceased or completed operations 1982-1990

Mineral Type	Number of sites	% of total
Chalk	5	1.2
China Clay	16	3.7
Clay/shale	12	2.8
Coal (deep mined)	1	0.2
Coal (opencast)	54	12.5
Gypsum/Anhydrite	1	0.2
Igneous Rock	21	4.9
Ironstone	1	0.2
Limestone	41	9.5
Oil/Gas (explor)	0	0.0
Oil/Gas (production)	0	0.0
Sand and Gravel	232	53.7
Sand (industrial/silica)	13	3.0
Sandstone	6	1.4
Slate	0	0.0
Vein minerals	28	6.5
Other minerals	1	0.2
Total	432	100.0

As may be expected the largest number of sites ceasing operation has been associated with sand and gravel. The relative proportion of sites by mineral type is broadly similar to the local authority survey, although deep mined coal is under-represented due to the absence of a postal reply from British Coal deep mines.

Restoration Performance by Operators

Differences of opinion were found between mineral operators and local authorities over restoration which was regarded as unsatisfactory. According to operators by 25 out of 432 sites (5.8%) were not restored to the authorities' satisfaction.

Operators generally accepted the need for modern conditions so that disturbance could be minimised and satisfactory restoration achieved. In practice, however, operators regarded many current conditions as illogical or badly worded and not capable of being enforced by the planning authority. Operators suggested that improvements in the standard of restoration in recent years reflected an increasing environmental awareness by the industry.

British Coal indicated that many of the problems with restoration on their deep-mined sites related to historic standards and methods of working, former General Development Order rights, and financial controls on the industry, rather than wilful default by the operator.

Overall operators suggested that the quality of restoration depended upon standards of working. A number of operators indicated that they had taken over older sites and faced problems in achieving a high standard of restoration arising from historic or inadequate methods of working.

The reasons stated for unsatisfactory restoration were wide-ranging. They are summarised in Table 2.8. The largest single category was slowness and related problems including difficulties encountered in making up shortage of fill materials. Reasons given within this category related to unforseen watertable problems, failure to agree a final restoration scheme within time limits and lack of continuity between local authority officers.

Failures of implementation of the planning system including unrealistic planning conditions, loosely worded planning conditions and lack of monitoring and advice from the local authority. Operators indicated that 70% of permissions completing or ceasing operations in the period 1982-1990 contained restoration and/or aftercare conditions.

Only 6.6% of operators thought there were likely to be problems in fulfilling current restoration and/or aftercare conditions.

Management of Restoration within Companies

A summary of the responsibility within the operating companies for ensuring that planning conditions for restoration are carried out is provided in Table 2.9.

In most operating companies it is the quarry manager or other production managers who are responsible and exert the strongest degree of control over the restoration process, particularly where restoration takes place progressively. However, a significant number of medium and larger operators have specialist restoration personnel to supervise works.

Table 2.8 : Reasons stated by operators for unsatisfactory restoration

Reason	Number of sites	% of total
Failure of implementation of the planning system	6	24.0
Local authority not satisfied with standard of restoration achieved	4	16.0
Activities suspended or possibility of reworking the site	5	20.0
Slowness or other problems including shortage of fill materials	9	36.0
Financial costs of the operations	1	4.0
Total	25	100

The introduction of specialist staff is a reflection of the increasing standards of restoration and its evolution from a 'materials handling' exercise to one of complex quality control involving careful management. A number of larger and medium sized operators in the survey urged that highly successful restoration could be achieved only by separating commercial responsibilities and pressures of production and marketing from restoration which needed to be regarded as a fixed and separate process.

Table 2.9 : Responsibility for ensuring that planning conditions for restoration and aftercare are carried out

Responsibility	% responsibility
Quarry managers at individual sites	40.0
Restoration manager covering all sites	16.0
Estates personnel	20.8
Others	23.2
Total	100.0

55% of operators in the postal survey indicated that they undertook some kind of training for those staff with restoration responsibilities. Most training has involved in-house courses although a number of respondents mentioned outside course including those organised by the Institute of Quarrying.

Some operators had also used landscape or other specialist consultants, although frequently this was at scheme preparation rather than implementation stage.

One major operator interviewed had undertaken an independent environmental audit of its sites. Others cited the SAGA restoration fund and BACMI code of practice as evidence of some operators willingness to achieve good practice.

Concern was expressed by a number of operators about the general failure of planning authorities to implement provisions leading to the updating of conditions under the 1981 Minerals Act, particularly as the Act offered the potential to control the activities of the irresponsible members of the industry.

Financial Provision for Restoration

Table 2.10 summarises operators policy on financial provision for restoration. Overall 34% of respondents indicated that they make no forward financial provision for restoration and aftercare, and that costs are incurred out of current revenue. However, as Table 2.10 indicates this experience divides sharply by type of operators, with over half of local operators making no forward provision but only 20% of national operators.

A number of operators argued that company taxation rules do not provide any incentives to make such provisions. The current tax position may be briefly summarised as follows:

- Financial provisions in advance of restoration costs being incurred is not tax deductible.

- Where restoration takes place progressively and the site is revenue earning, restoration expenses are tax deductible in full in the year in which such outlays are incurred.

- Where restoration takes place at the end of the site's economic life, restoration expenses are tax deductible against revenue in the year in which they are incurred even though this revenue is generated on a different site within the company.

- Where the company has ceased mining operations when restoration is due to take place, only costs that may be incurred in the following three years are deductible against the profits of the trading years.

Table 2.10 : Policy on financial provision for ensuring compliance with site restoration and aftercare conditions

Type of Provision	% of operating companies	% of national operators	% of local operators
General provision is made in advance covering all company sites	6.8	5.1	8.0
Specific provision is made in advance for each site	9.7	12.8	8.0
Full provision is made when site development commences	2.9	5.1	2.0
Provision is accumulated on an incremental basis as sites are developed	34.0	48.7	28.0
No forward provision is made restoration aftercare is incurred out of current revenue	34.0	20.5	52.0
Other approaches	12.6	7.8	2.0
	100	100	100

Thus, large companies with several sites or with sites with progressive restoration schemes, may take full advantage of tax deductibility. The tax system appears to penalise smaller operators, and to discourage making financial provision.

If the tax system were to be changed to give relief on provision in advance of restoration costs some mechanism would be required to ensure that the "fund" would be spent on restoration, eg. held by a third party on in trust.

Operators argue that the present tax position is especially unsatisfactory for companies operating from only one site in relation to expenditure incurred at the end of working or after cessation of workings. While expenditure up to three years after working qualifies for relief, restoration of a large site could take longer than three years. Tax relief would also not be available to cover an additional aftercare period although these costs are usually a small proportion of overall costs and often smaller than restoration costs. Moreover, as the expenditure for taxation purposes is deemed to be incurred on the last day of trading, relief could again be lost if insufficient profits were made in that tax period.

Generally operators suggested that a tax system which encouraged forward provision would permit a portion of the estimated total costs of restoration to be tax deductible against assessed income of an extraction operation in each year in which the mining operations are conducted.

Specific Restoration Problems by Mineral Type - borrow pits, coal, metalliferous sites

The local authorities' interviewed in the surveys were largely of the opinion that three factors were principally related to failure to restore to satisfactory standards : older permissions; methods used by operators for workings sites; the technical standards and competence of the operator. Mineral type was regarded as far less significant than these, except that there is a correlation between mineral type and the age of permissions.

However, mineral types that do have some important distinguishing differences are: borrow pits, where the main problems are the short-term nature of operations and non-competence of operators at restoration; vein minerals, because of pollution risks; some private coal sites, where reworking of old sites means high site restoration costs relative to the value of the current output.

Borrow pits are temporary extraction sites on, or in the vicinity of, major civil engineering construction sites which are used solely to supply aggregates for the construction project, and which are sometimes used for the disposal of surplus materials from the construction sites. Permissions are usually granted to meet a particular peak in local demand. Planning authority experience indicates that they are often sought and operated by small scale civil engineering sub-contractors who have little experience or interest in restoring mineral sites to the high standards required. Given the short time period of operation, experience suggests enforcement action is generally unrealistic.

Within the private coal sector, problems primarily relate to the smaller peripheral operations including tip reworking and private deep mining where the scale of restoration may be large in relation to the value of minerals extracted. Many of the operators within this sector are small and have limited resources to meet any unforseen technical failure.

Coal extracted by British Coal presents special difficulties, already referred to, in relation to historical workings and past GDO procedures. Authorities were concerned about the prospects for restoration of tipping sites for colliery spoil or reworking these materials should privatisation proceed. Many sites do not have restoration conditions, and would not present a commercial proposition if they did.

British Coal have accepted a responsibility for restoration of collieries and tips which close in the four years commencing 1 April 1990. The Agreement applies to any deep-mine started before 1 July 1948 which closes within four years commencing 1 April 1990. Sites will be restored to any amenity use to a standard agreed with the mineral planning authority. Where planning consent provides that the site is appropriate for development, British Coal will contribute to any subsequent Derelict Land Grant aided scheme the notional cost of restoring the site to amenity use to a standard acceptable to the mineral planning authority.

Surface deposits of coarse and fine waste materials from the working of base metals and associated minerals (ie. vein minerals) come within the remit of planning control but this is not the case for outflows and overflows to surface waters of water from abandoned and disused mines. Both can cause pollution problems. The wastes, if left bare, may contain elements which are toxic to or affect plant growth or are harmful to grazing animals. Planning conditions for restoration and aftercare may make the affected areas suitable for an after-use.

However, pollution from mine water, of which Wheal Jane in Cornwall is a recent and serious example has raised a number of problems. Such discharges fall within the Water Resources Act, 1991 for any remedy but currently no offence is committed where water from an abandoned mine enters "controlled waters".

2.2 Assessment of the Effectiveness of Existing Procedures

Perceptions by local authorities and operators of the success of existing arrangements in achieving satisfactory restoration are different. Operators assessment of the number of sites failing to achieve satisfactory standards is far less than assessments made by local authorities.

Operators interviewed appreciate the need for restoration conditions but are frustrated by the limitations, as they perceive them, of the planning authorities in implementing planning policies.

The survey findings lead to the conclusion that according to local authorities the extent of failure to restore to satisfactory standards is significant. Of sites ceasing operations in the period 1982-90, 27% were not restored to a satisfactory standard or were not restored at all. Factors contributing mainly to satisfactory restoration were considered by the local authorities to be technical competence of the operator in undertaking the restoration scheme, methods used to work the site and the existence and specification of the restoration conditions of the planning permission.

The evidence suggests that the proportion of sites failing to be restored to satisfactory standards has changed little since the time of Stevens report, although it remains unclear as to the extent to which there has been a change in the standards being used to judge what is satisfactory. However, the mix of reasons associated with failure to restore has changed since Stevens with a smaller proportion of sites failing due to problems with the implementation of the planning system and a greater proportion failing due to diverse technical problems. Table 2.11 summarises the reasons for failure from the current survey and Stevens survey.

Of the principal reasons for failure, the largest group consists of diverse technical and other problems linked to inadequate practices by mineral operators or methods of working, lack of expertise and, in some instances, circumstances beyond the operators' control. Although presented as one category of failure, the detailed reasons were found to be highly diverse.

Financial failure of operators represents a small proportion of the total number of sites and this is a finding which is consistent with the earlier evidence from Stevens. In the current survey, only 5% of sites were restoration was not to a satisfactory standard could be attributed to the financial failure of the operator. This represents about 1.6% of all sites were operations ceased in the period 1982-90.

Limitations of the implementation of the planning system constitute about 35% of sites where there was a failure to restore to satisfactory standards. Reasons may be attributed to the absence or inadequacy of conditions of planning permissions, failure to monitor operators' activities properly, or failure to enforce operators to comply with planning conditions.
As regards to the absence or inadequacy of conditions, the problems arise from historical permissions which do not meet contemporary standards. Current practice is generally regarded

as satisfactory. Attempts to implement the provisions of the 1981 Minerals Act, and in particular to modernise permissions, have not been successful on the whole, due mainly to limitations faced by local authorities of staffing and manpower, and only incidently a matter of expertise.

Table 2.11 : Summary of reasons for failure[1]

	% of Sites Failing	
	Current Survey	Stevens Survey
Failure of implementation of the planning system[2]	34.6%	57.2%
Financial failure of operator	5.0%	3.3%
Diverse technical problems	42.3%	15.1%
Others[3]	18.0%	24.4%
	100	100

Notes (1) : Sites failing to achieve satisfactory standards of restoration
(2) : includes unrealistic planning conditions, loosely worded planning conditions, no requirement to restore and lack of monitoring
(3) : Includes: slowness or delays; ownership changes or problems.

Similar problems of resources, rather than powers, have influenced the extent to which authorities engage in monitoring. On the whole, the minerals planning departments are well organised to undertake monitoring activity but find that the need for intensive monitoring at sensitive times in the workings of sites causes undue strain.

Similarly with enforcement, where the powers as enhanced by the 1991 Act are generally regarded as sufficient for the principal purposes. Complexity, delay and the resource effort needed to undertake successful enforcement action acts as a deterrence and relatively little action is taken by mineral planning authorities to achieve satisfactory restoration through the enforcement procedures.

Thus, the limitations of the planning system have less to do with the powers available to local authorities at the current time than their resources to implement the system.

There is a reluctance on the part of mineral planning authorities to use the enforcement mechanism because of the legalistic and lengthy processes involved. Some authorities rarely used stop notices because of the fear that compensation would be payable if the notice was subsequently quashed on grounds other than the grant of permission by the Secretary of State. The 1991 Act introduced the breach of conditions notice which authorities considered would improve their enforcement powers.

There are some special problems not adequately dealt with by the existing arrangements. Borrow pits cause local authorities particular difficulties as a consequence of their short-term nature and the lack of competence to restore by some operators. Private coal sites in certain circumstances also present difficulties, often because the sites are old and are being reworked with the effect that the value of the output is low relative to the magnitude of the restoration costs.

Some authorities expressed concern over restoration of some British Coal sites in the event that privatisation proceeds, given the absence of restoration conditions in relation to reworking of tips on many of these sites and the lack of economic incentive to acquire them should restoration conditions be imposed.

There is little evidence to suggest that the system of company taxation operates to frustrate successful restoration. Nevertheless, it provides for no incentives to operators, particularly the smaller ones, to make provisions for restoration despite the clear desirability that they should do so.

In the postal survey of local authorities, and in the follow-up interviews that took place, enquiries were made about policies towards restoration bonds. There was general support for a bonding mechanism amongst planning authorities although no consistency as to the form it should take nor was there a clear view about the objectives to be served. This matter is dealt with, as part of a general review of experience of financial guarantee schemes, in the next chapter.

3 EXPERIENCE OF FINANCIAL GUARANTEE SCHEMES

The following are areas of relevant experience with financial guarantees and are reviewed in this chapter:

- Restoration guarantees/bonds in Britain under voluntary Agreements under planning (or local government) acts.

- Restoration guarantees/bonds for private coal working under local acts for West Glamorgan, Mid Glamorgan and Dyfed.

- Industry schemes covering the Sand and Gravel Association Restoration Guarantee Fund and British Coal Opencast bonding arrangements on licensed sites.

- Ironstone Restoration Fund.

- Indemnities, guarantees and performance bonds used in contracts, such as those for highways and construction projects.

- Overseas experience.

Each of these six areas are considered from the point of view of insights that it provides for the introduction of a possible scheme for mineral restoration and aftercare financial guarantees in Britain.

3.1 Voluntary Agreements

A summary of restoration bonds held by planning authorities responding to the postal questionnaire is provided in Table 3.1. This indicates the existence of 61 bonds among the 52% of responding planning authorities. Grossing up, this equates to a national estimate of approximately 100-120 bonds or 2% of all current working sites. From the current survey 77% of these bonds are held in respect of opencast coal sites. 67% of bonds are held on Scottish sites.

These bonds have mostly been obtained through voluntary Agreements under the planning or local government acts. Other authority experience has included bonds negotiated in respect of planning permissions for waste disposal operations associated with mineral working.

Typically, when reaching Agreements the planning authority obtains an understanding with the operator as to the expected scale of restoration costs and aftercare in relation to a scheme for the site, calling on the advice of the engineering department of the authority, MAFF and other sources. Once the sum has been agreed it has generally been left to the operator to decide the form that the bond to cover it should take. Bond sums have varied in value between £3,000 and £900,000, depending on the nature of the operations and type of restoration required and typically, sums have been in the £50,000 - £100,000 range.

Table 3.1 : Restoration bonds held by planning authorities under voluntary Agreements or other arrangements

Mineral Type	Number of Bonds[1]	% by Mineral Type
Chalk	1	1.6
Clay/Shale	1	1.6
Coal (deep mined)[2]	1	1.6
Coal (opencast)[2]	47	77.0
Sand and Gravel (construction)	6	9.8
Sandstone	1	1.6
Vein Minerals	3	4.9
Others	1	1.6
Total	61	100

Notes : (1) Based on the responses of mineral planning authorities in England and Wales and planning authorities responsible for minerals in Scotland to the postal questionnaire.
(2) Excludes bonds held by South Wales authorities under 1987 Local Act powers.

In the few cases of very large operators having agreed to a financial guarantee, no substantial difficulty has been encountered in obtaining a guarantee bond from the operators bank. Smaller operators have approached banks and insurance companies but have encountered greater difficulties in securing guarantee-type arrangements.

For example, a case was provided where a landowner sought planning permission for a sand and gravel operation prior to reaching a lease agreement with a large operator. In this case, the landowner approached the insurance industry for a bond likely to be required over 20 years and for a site which would generate £20 million output. Despite the landowner being established in the community and having extensive assets including the potential asset of the mineral in the ground, this landowner had been unable to obtain a bond within six months of starting a search for one. The matter has now been settled in principle, with the local authority agreeing to a 3-year bond, annually renewable.

Where small operators have been unable to obtain bonds the security has had to be provided in other ways. One example was of a small operator working sand and gravel pits in the South of England with a turnover of about £1 million per annum. This operator was of good general standing in the local community and had a good restoration record in the industry. He had difficulty with obtaining a bank bond, and cash security of £50,000 had been lodged in a bank account to be released in the event of failure to restore, and with an additional guarantee secured against the operator's house.

Box 3 provides an example of the wording of a financial guarantee by Agreement.

Planning authorities argue strongly that these bonds have undoubtedly provided a strong incentive for operators to complete restoration works. It has not been possible during the course of the study to identify cases where these bonds have been forfeited.

Box 3 : Example of a financial guarantee by Agreement under the planning acts

AGREEMENT
between
_____ Council
and
_____ Company
and
_____ Bank
[under S106 of the Town and Country Planning Act 1990]

NOW IT IS AGREED AS FOLLOWS:-

1. In as much as the permission will oblige the Owners to progressively and continuously reinstate and to finally restore the said subjects in accordance with the terms of the permission, the Owners hereby undertake and agree to observe, perform and fully implement the obligations imposed by the permission.

2. If the Owners shall at any time fail to perform and observe any of the requirements of the permission or of any such future planning permission or of Clause 3 hereof in so far as relating to reinstatement and restoration of the subjects and such failure shall continue unremedied after Twenty eight days' notice thereof has been given by the Council to the Owners, or if the Owners otherwise fail to proceed with reinstatement or restoration works expeditiously or otherwise in a manner satisfactory to the Council's Director of Planning, or if the Owners shall go into receivership or liquidation (except liquidation for the purpose of reconstruction or amalgamation) then, without prejudice to the right of the Council to exercise any statutory powers, the Owners will pay to the Council within Fourteen days of receipt of written notice from the Council such sum of money as the Council estimate will require to be spent by the Council to remedy the Owner's failure and will grant the Council, their employees, agents and contractors all necessary right and authority to enter upon the said subjects and to perform the work necessary to remedy the Owners' default. On completion of the works by the Council the Owners will pay the Council on demand any further sum by which the sum expended by the Council exceeds the estimated sum.

3.(a) In security of performance by the Owners of their obligations as contained in and referred to in Clause 4 hereof, the Owners will provide to the Council at the Owners' expense a bond or financial guarantee of performance of the Owners' obligations, such bond or guarantee to be by a bank or other financial institution or company approved by the Council and for an amount and in a form acceptable to the Council.

(b) The initial amount of such bond or financial guarantee shall be the sum of   POUNDS (£   ) and shall be reviewed by the Council prior to the second anniversary or the last date of execution of this Agreement, or within Fourteen days of written notification by the Council to the Owners of the result of each such review, whichever shall be the later. Such reviews shall take account of the cost of implementing the reinstatement obligations contained in the permission, all as approved by the Council in terms of the permission or any such future planning permission

(c) The Owners will provide, or will ensure the provision, to the Council each year of a copy of the Annual Report and Accounts of the body which provides the said bond or guarantee.

4. This Agreement shall continue in force until the whole of the reinstatement and restoration obligations contained in the permission and any such future planning permission shall have been performed, except as the Agreement may have been previously discharged in whole or in part by the Council.

5. No compensation shall be payable by the Council to the Owners in respect of or arising out of this Agreement.

6. The parties consent to registration hereof for preservation and execution:

[Signatures]

3.2 South Wales Local Acts

In South Wales the counties of West Glamorgan, Mid Glamorgan and Dyfed have powers under local acts to require bonds for private coal operations. Although the current provisions were granted in 1987, within Mid Glamorgan provisions have existed since 1973, and reflect the problems experienced by the former Glamorgan County Council in the 1950s and 1960s with mineral operators going into liquidation and leaving mineral sites unrestored.

The wording of the 1987 Acts are the same for the three counties and provides that the applicant or operator of the site should provide security for the performance of any conditions subject to which the planning permission is granted relating to restoration or aftercare. The Acts provide that this security should be provided by a bond or by such other means as may be approved by the mineral planning authority. Relevant wording is provided in Box 4.

A number of bonds and other forms of financial security have been provided, with Dyfed and Mid Glamorgan holding approximately 20 bonds each and West Glamorgan holding over 100 bonds.

The amounts covered by the bonds vary. Most of the licensed deep mine bonds are for amounts of between £5,000 and £50,000. In the case of opencast and tip reworking schemes the figures have been larger and up to £400,000 with one exceptional case of £1.1 million. Most are typically in the range of £50,000-£100,000.

For the private deep-mined sites, the principal items for restoration are audit entry, on-site loading platform and removal of temporary plant, which accounts for the modest size of bonds. For the opencast sites, the restoration mainly consists of backfilling and treatment.

The arrangements for securing bonds have progressed beyond some initial difficulties to the point where they seem to have become an established and accepted routine. Most bonds are secured with local branches of the major high street banks, although a few are with specialist insurance or finance companies. Typically, the bank will make an annual charge of between 2 and 3% per annum of bond amounts for acting as guarantor. Having built up local experience, few operators have found difficulties in securing bonds, although bond sums have generally been small for the smaller operators.

Many of the sites where bonds have been imposed are still operational. In eight cases where operations have ceased the local authority concerned has called the financial guarantee. In seven of these, where bonds had been deposited, there was financial failure of the operator. In the eighth, where the operator had lodged cash rather than a bond, the operator chose to abandon the site, forfeit his money and leave restoration to the authority.

Wording of bonds varies but an example is provided in Box 5. The bond can be called when the operator is "in default". However, independent legal opinion, which is supported by views and experience of guarantors, suggests that the word "default" would, in practice, need to be given very precise meaning that could only be established through legal process. Financial failure as a reason for not meeting restoration obligations qualifies as default, however.

Box 4 : Extract from South Wales local act

6   c.vii   Mid Glamorgan County Council Act 1987

Coal-mining operations performance bonds

6.—(1) Planning permission under the Act of 1971 for development in the county which consists of or includes the winning and working of coal (including the extraction of coal from a mineral-working deposit), or other operations in, on, over or under land for or in connection with such winning and working of coal, may be granted subject to the requirement that any of the following persons, other than the National Coal Board, namely:-

(a) the applicant for the planning permission, or, as the case may be, the appellant against an enforcement notice; or

(b) any other person who carries out the development authorised by the planning permission;

shall provide, to the satisfaction of the mineral planning authority, before commencing, or, as the case may be, proceeding with, the development, security for the performance of any conditions subject to which the planning permission is granted relating to landscaping or the preservation, restoration or reinstatement of the land forming the site of the development, including any restoration condition or aftercare condition.

PART II —cont.

(2) The security for the performance of conditions of a planning permission which may be required under subsection (1) above may be provided____

(a) by a bond, guaranteed by a guarantor approved by the mineral planning authority, for payment to the mineral planning authority, in default of such performance, of such a sum as may be required to secure compliance with those conditions; or

(b) by such other means as may be approved by the mineral planning authority to secure compliance with those conditions.

(3) Where the costs incurred by the mineral planning authority in carrying out works or operations to perform, or complete the performance of, conditions of a planning permission for which security is required under subsection (1) above, exceed the sum for which security is provided in accordance with the foregoing provisions of this section, the mineral planning authority may recover the excess from the persons specified in subsection (1) (a) or (b) above.

(4) Where in pursuance of the foregoing provisions of this section a person is required to provide security for the performance of conditions of a planning permission and before the completion of the development authorised by that planning permission the right to carry out that development becomes vested in some other person, then ____

(a) upon being notified by the National Coal Board for the purposes of this subsection that the right to carry out that development has become vested in it; or

(b) upon being satisfied that some other person in whom for the time being the right to carry out that development is vested has provided security for the performance of conditions of the planning permission.

the mineral planning authority shall, in either case, upon the request of the person who, before such vesting had been required to provide security for the performance of conditions of the planning permission, release that last-mentioned person from all liability in respect of the security provided by him.

(5) This section shall not apply to any planning permission granted after the coming into operation in the county of any other enactment providing means to secure the performance of such conditions as are mentioned in subsection (1) above.

(6) In this section ____

(a) "coal" means bituminous coal, cannel-coal or anthracite; and

(b) expressions to which meanings are assigned by the Act of 1971 have the same respective meanings.

> **Box 5 : Example bond wording - South Wales**
>
> Dated _____
>
> _____ mineral operators Ltd
>
>   and
>
> _____ Bank PLC
>
>   to
>
> _____ County Council
>
>   BOND
>
> in respect of land at _____
>
> in the county of _____
>
> **KNOW ALL MEN BY THESE PRESENTS** that _____ of _____ and _____ BANK PLC (hereinafter referred to as "the Surety") are pursuant to the provisions of Section 6 of the _____ County Council Act 1987 bound to the ___ _____ COUNTY COUNCIL (hereinafter called "the County Council") in the sum of _____ Thousand Pounds (£_____) to be paid by the Developer to the County Council for which payment we bind ourselves and each of us jointly and severally and the successors and assigns of each of us the Developer and Surety by the presents
>
> **SEALED** with our respective seals this 1st day of November One Thousand Nine Hundred and Ninety One
>
> **WHEREAS** it is a condition to a planning permission (hereinafter referred to as "the Planning Permission") bearing the code XX/89/XXXX granted by the County Council in respect of land in the County of _____ requires the execution and performance of the restoration and aftercare works referred to in the conditions annexed to the Planning Permission
>
> **NOW THE CONDITION OF THE ABOVE WRITTEN BOND** or obligation is such that if the Developer shall well and truly perform fulfil and keep condition then the above written Bond or obligation shall be void but otherwise the same shall remain in full force and virtue
>
> **IN WITNESS** whereof the Surety has caused its Common Seal and the Developer has set his hand and seal the day and year first before written
>
> THE HAND AND COMMON SEAL OF XXXX
>
> was hereunto affixed in the
>
> presence of:-
>
> THE COMMON SEAL OF XXXX BANK PLC
>
> was hereunto affixed in the
>
> presence of:-

As in the case of bonds under voluntary Agreements, the relevant planning authorities argue strongly that these bonds have provided a strong incentive for operators to complete restoration works to a satisfactory standard without having recourse to enforcement procedures.

3.3 Industry Schemes

SAGA

The Sand and Gravel Association (SAGA) has a restoration guarantee fund for its members. This scheme is operated as a separate limited company, the principal objective of which is as follows:

> "To provide an indemnity to MPAs in respect of members of the Fund against costs occasioned by the failure of the member to restore any land in which he has an interest and used by him for the purposes of extracting aggregates to the state required by a planning condition".

The Fund was set up in 1975. Operators' contributions are single, initial payments reflecting the size of the firm's output. There is the scope for an "extraordinary call" on other members if an operator defaults and the whole Fund is absorbed by subsequent restoration costs. This facility allows SAGA to call for fresh funds from members. The Fund is relatively modest in size (about £200,000) and is invested securely. SAGA undertakes careful vetting of new members for there is the possibility of a demand to join simply in order to reap the benefits of membership of the Fund for companies that are having difficulty securing planning consents without financial guarantees. The Fund has never been drawn on.

To claim against the SAGA fund an authority must meet the following conditions:

> "1 The planning or other appropriate Authority has used every endeavour to secure compliance by the member with the said condition, including preventing the member from continuing work on the land in question and the taking of enforcement proceedings against him.
>
> 2 The member shall have failed, by reason of insolvency, to comply with the said condition for the period of one year from the conclusion of enforcement proceedings.
>
> 3 The member, at the time of commencing to extract ballast from the land in question, was lawfully entitled to do so by virtue of his ownership of an estate or interest in the land or a licence granted to him.
>
> 4 Work shall not have ceased on the land in question by the said member before 1st January, 1974".

Thus, the Fund is in effect to be called only in the event of financial failure.

It is difficult to evaluate the SAGA restoration guarantee fund through lack of experience with it being called. In the past, some local authorities have been reluctant to accept the scheme as sufficient security and have sought bonds as part of voluntary Agreements with SAGA members. SAGA has vigorously supported its members who have been confronted by authorities additionally seeking bonds.

Separate from the Fund is a SAGA Restoration Award scheme where a panel of judges looks for examples of restoration often which go beyond the requirements of planning conditions. This Award scheme helps promote the concept of SAGA members being responsible operators, as well as raising the profile of the SAGA Fund.

British Coal Opencast Executive

British Coal Opencast Executive has bonding arrangements but these are somewhat different from those which would be entered into between local authorities and operators. Contrary to common impressions, they are not for restoration. These bonds are to cover British Coal's royalty payments on licenced sites in the event of an operator no longer being able to work the site eg. through liquidation or walking off the site, and as a protection against losses of royalty from unauthorised extraction. Liability for restoration is borne by the contractors who operate the sites.

In terms of procedures, because of the short time periods of operation (2-5 years) and the credit-worthiness of most of the operators almost all the British Coal securities are of a guarantee-type. These are generally obtained from major clearing banks or insurance companies. British Coal has no formal list of acceptable institutions because mostly bonds come from well known sources. Historically a number of securities were cash deposits, known as security funds. Although British Coal still holds a few of these, there are now few operators who are unable to provide a guarantee-type bond.

3.4 Ironstone Restoration Fund

The Ironstone Restoration Fund was set up under the Mineral Workings Act 1951 for the purpose of financing the restoration of land in England following opencast ironstone operations. When the Fund was established there was a backlog of dereliction from past ironstone workings. There were many small and often irresponsible operators, and restoration performance was poor at a time when extraction methods were changing in the direction of making the restoration tasks even more difficult. The Fund was operated by Government. Contributions to the Fund were received from operators (mainly) and Government. Sums were paid out to operators, local authorities or others for restoration works. The standard of restoration to agriculture or forestry use was assessed by MAFF and the Forestry Commission respectively.

Contributions payable by operators were linked to output, measured by tonne of ironstone extracted. The initial levy on operators was 3d per ton. Under the formula laid down for restoration, operators contributed a "standard rate" of £110 per acre towards the cost of restoration whilst the Fund met the rest; it had, therefore, a topping-up feature.

Substantial restoration took place after 1951. Operators exploited and restored over 1,440 ha of land, whilst 560 ha of derelict land in 1951 was also restored using the Fund. From this standpoint of physical restoration the Fund was generally regarded as a success.

A serious defect of the Fund was its failure to make allowance in its levies for changing costs. The Fund paid an increasing proportion of total costs as a result of the general inflation of costs. There was, additionally, a decline in iron-ore output which caused income to the Fund to fall. In response to these conditions, the Government belatedly took action in 1971 to enable changes to be made to both a contributions and the "standard rate" of restoration. This helped, but could not remove the underlying flaw which was that restoration costs from past workings were a liability at a time when the economic value of the remaining mineral was tending to zero. The Fund was a substantial financial failure at the time of wind-up in 1985.

> **Box 6 : Specimen form of on demand bank guarantee**
>
> .... Bank PLC
>
> BONDING AGREEMENT
>
> In consideration of your entering into an agreement with .... ("the Customer") for (inter alia) .... ("the Agreement") we, ... Bank Plc hereby undertake to pay you on receipt of your first demand on us in writing stating that the Customer has become liable and has failed to pay you sums due under or by reason of their breach of Agreement, the sum stated in such demand.
>
> PROVIDED THAT:-
>
> 1. This undertaking is personal to you and is non-assignable.
>
> 2. Our liability hereunder shall be limited to a sum or sums not exceeding in aggregate .... and shall expire on .... except in respect of any written demand for payment complying with all the requirements hereof received by us at this office on or before .... after which date this undertaking shall be void whether returned to us or not.
>
> 3. Any demand hereunder must comply with all the above requirements and the signature(s) thereon must be confirmed by your bankers.
>
> 4. This Agreement shall be governed and construed according to English Law.
>
> Dated this .... day of .... 19 ....
>
> Per Pro .... Bank Plc
>
> Duly authorised by resolution of the Board.

There were, additionally, various problems with this Fund including arguments over the formula to be used in compensating landowners for their efforts, disputes over individual claims and problems of administration.

3.5 Indemnities, Guarantees and Performance Bonds under Contracts

Arrangements whereby one person agrees to be answerable for the liability of another to a third person are, of course, common in commerce and business generally. Such arrangements are invariably under contract, and hence differ in principle, although possibly not in substance, from the security that might be given under the process of regulation as operated by the planning acts. The surety (ie. bank or insurance company) undertakes to be responsible for the whole or part of the possible debt or obligation of another (the debtor or principal, ie. the mineral operator) to a third person (the creditor ie. the local authority).

There is extensive case law in relation to contracts of **guarantee and indemnity** but some general observations on the way these contracts work may be helpful in the context of financial guarantees for mineral workings:

- Guarantees may be made retrospective, covering an already existing liability but there are considerable problems with this, in particular the consideration for it and most guarantees cover a future liability.

- Guarantees may be made specific, that is one which covers a specified liability, or continuing which covers liabilities outstanding from time to time.

- For continuing guarantees, it is common for the liability to be strictly limited either to part of the liability or, alternatively, in overall amount.

- It is common for the guarantee to be conditional on the occurrence of some event.

A surety may be discharged through the inequitable conduct of the creditor. In the event of a local authority acting unreasonably, or changing the characteristics of a planning condition, there may be problems in substantiating that a financial guarantee undertaken by the mineral operator would still be valid.

Performance bonds are simple covenants by one person to pay another. They may be subject to conditions. Performance bonds are often called guarantees but in fact they are not, in the true sense of the term. They are simply undertakings to pay a certain sum on the happening of a contingency, namely the failure to perform a contract. They are not guarantees for the performance of the contract itself.

Performance bonds are of major importance in relation to the construction industry. The employer is concerned that the contractor or builder will perform the works that he has undertaken to do or, if he does not, that at least the employer will be compensated by some cash payment in default. The contract between the employer and the builder typically requires that the contractor procure that his bank gives an undertaking to pay sums of money if the contracts are not performed. The bank then provides the bond for payment which is the performance bond, which is a certain sum on the happening of a contingency, namely the failure to perform the contract.

As a condition of giving a performance bond, the bank will invariably require a counter-indemnity to be given to it by the contractor. Thus, if the bank does pay under the performance bond, the contractor will in turn have to repay the bank. Accordingly, the supplier or contractor will be concerned that the bank is not bound to pay under the bond saving circumstances where it is fair that the bank should do so. Bonds fall into two principal categories:

- "On-demand" bonds - which are unconditional obliging the surety or bondsman to pay simply on-demand by the creditor. In these cases, the surety will have obtained the counter-indemnity in advance. An example is provided in Box 6.

- Conditional bonds - in this case the employer has to establish damages occasioned by the breach of the conditions to the contract and, if he succeeds, he recovers the amount of the damages proved. In these cases the beneficiary of the bond, such as the local authority, has to prove loss. An example is provided in Box 7.

Box 7 : Example of insurance company bond

INSURANCE COMPANY

BOND No.

BY THIS BOND We
whose registered office is at

(hereafter called "the Contractor")

and   Insurance Company
whose office is at

(hereafter called "the Surety") are

held and firmly bound unto

(hereafter called "the Employer")

in the sum of

for the payment of which sum the Contractor and Surety bind themselves their successors and assigns jointly and severally by these presents.

Sealed with our respective seals and dated this
day of   19

WHEREAS the Contractor by an Agreement
made between the Employer of the one part and the Contractor of the other part has entered into a Contract (hereinafter called "the said Contract") for the construction completion and maintenance of certain Works as therein mentioned in conformity with the provisions of the said Contract.

NOW THE CONDITION of the above-written Bond is such that if the Contractor shall duly perform and observe all the terms provisions conditions and stipulations of the said Contract on the Contractor's part to be performed and observed according to the true purport intent and meaning thereof or if on default by the Contractor the Surety shall satisfy and discharge the damages sustained by the Employer thereby up to the amount of the above-written Bond then this obligation shall be null and void but otherwise shall be and remain in full force and effect but no alteration in terms of the said Contract made by agreement between the Employer and the Contractor or in the extent or nature of the Works to be constructed completed and maintained thereunder and no allowance of time by the Employer or the Architect under the said Contract nor any forbearance or forgiveness in or in respect of any matter or thing concerning the said Contract on the part of the Employer or the said Architect shall in any way release the Surety from any liability under the above-written Bond.

The Common Seal of

was hereunto affixed
in the presence of:

DIRECTOR/SECRETARY

The Common Seal of
        Insurance Company
was hereunto affixed
in the presence of:

ATTORNEY-IN-FACT

In construction projects, bonds normally are only called in cases of liquidation even though the bonds themselves may be couched in wider terms. Since the bonds cover contractual arrangements, in the event that the work is defective the employer would simply fail to pay the contractor for work done. The standard civil engineering bond is conditional, and it is for this reason that it is usually only called in cases of financial failure.

3.6 Overseas Experience

Other countries have acted to control the environmental degradation caused by mineral working.

Approaches to control have fallen into two basic types. The first includes those countries which have long-established planning systems, broadly similar to the UK, which have also made legal provision for financial guarantees as a supplementary power but as yet in limited usage. The second includes those having arrangements developed in recent years and often including detailed performance standards linked to financial guarantee or bond requirements. As in the UK, systems of control attempt to place primary responsibility on the mineral operator to restore sites.

This section considers the experience in a selection of these countries.

Germany

Primary responsibility for restoration policy resides with the Länder who have adopted differing systems of control. Among the differing regulatory systems, North Rhine-Westphalia is widely praised for its environmental results.

The North Rhine-Westphalian regulatory system involves a three-tiered administrative structure. The top level comprises the Minister of Economics, Resources and Transportation who is ultimately responsible for the regulation of mining. The second administrative tier is the chief State mining authority which has responsibility for mining and sets standards and regulations governing land reclamation and guidelines on the management and cultivation of soil. The third tier is the Land mining authority. This authority plays the most important role of reviewing and approving mineral extraction 'plans', which include schemes of restoration, and inspecting and enforcing these extraction plans.

Under North Rhine-Westphalian law, a mine plan is deemed automatically approved unless an objection is raised within two weeks. Consequently the land mining authority routinely raises objections, and review of the mine plan is conducted over a period of between six months and two years, the actual duration depending upon the complexity of the plan. Other government agencies, such as the Forestry Office, are consulted during the review process with changes being negotiated with the operator.

State mining authority officials argue that the long time period spent reviewing the restoration plan means that most operators comply with the plan and restoration has consequently been successful. The system is backed up by regular monitoring. Any problems are usually handled by discussion and informal agreements, without the use of formal enforcement. In other circumstances the threat of formal action has proved sufficient.

Although no Federal legislation exists for the purpose, some mineral operators are encouraged to provide financial guarantees to ensure restoration. The operator suggests the sum when applying for permission and may also suggest the type of guarantee. However, many operators within the minerals industry are large (particularly coal operators) and bonds are considered unnecessary because of the low risks of default. Restoration is usually achieved through implementation of the mining plan by the operator.

France

In France the primary legislation controlling mineral working is the Décret No. 79-1108 of 20 December 1979 containing 46 detailed articles covering the definition of minerals covered by the Décret, the application procedures, information required in support of an application and provision for restoration. Mineral extraction is also subject to policy controls within the Code Minier (the mining code) and the relevant Plan d'Occupation de Sol (the 'local plan').

With respect to restoration and aftercare provisions, the Décret requires applicants to submit details of the soils and overburden as well as proposals for their conservation and storage. Applicants must also submit detailed working restoration and aftercare plans which will include the steps necessary to achieve the proposed afteruse. Applicants must also provide a statement of their financial and technical capability.

Primary reliance is placed on the operator to implement the proposals. However, in the event of default a stringent system of enforcement exists which will require the operator to rectify the default and will also involve notification of the breach to the Minister in charge of mines, the Industry Minister and the Mayors of other Communes.

Powers exist to require bonds or financial guarantees. In practice, these have been rarely used reflecting the relative success of existing mechanisms.

Denmark

The Raw Materials Act 1977 is the principal legislation controlling surface mineral working. This covers most minerals including rock, sand, gravel, clay, limestone, chalk, peat and similar deposits. The scope of the Act is to ensure that exploitation is based on the best balance of economic, social and environmental costs.

The laws require mineral companies to restore worked out sites through a series of conditions attached to permits to be implemented by the operator and which may be enforced in the event of default occurring. Operators may also have to provide a bank guarantee to the county authorities.

Portugal

Recent legislation in Portugal including the Decree-Law 88/90 of 1990 has included provisions for the control of mineral working and restoration which includes powers for bonding. The Decree defines restoration to include construction of facilities which as far as possible fit in with the surrounding countryside and, on completion of exploitation, reconstitution of the terrain for use according to the purposes for which it was employed prior to commencement unless some other use has been determined in a plan approved by the competent authorities.

In order to gain an exploitation 'concession' operators must enter into a contract with the State and obtain a licence for ancillary operations. The details bound by the contract include the extraction areas, the type of minerals, duration and restoration conditions. The Act also makes special provision for the separate storage of topsoil with a view to the subsequent reconstitution of terrain and flora. The bond must be lodged with the State to cover restoration liabilities. The time-period since this legislation was introduced has not permitted evaluation of its effects.

As with Portugal, all mineral workings in Spain (above a certain minimum size) must have lodged a financial guarantee to cover site restoration.

USA

The major federal legislative effort to control mineral working has been the Surface Mining Control and Reclamation Act 1977 which is limited to coal. The aim of this Act was to "establish a nationwide programme to protect society and the environment from the adverse effects of surface coal mining operations". It provides for reclamation bonds.

The performance standards established by SMCRA require that the operator restores the affected land to a condition capable of supporting the use which it was capable of supporting before mining, or to a higher or better use. Amongst other provisions the operator must also restore the approximate original contour of the land by backfilling, grading and compacting; minimise disturbances to the hydrologic balance; conduct reclamation as quickly as practicable with the extraction operation and establish a permanent vegetative cover in the affected area. There are also special performance standards for particularly vulnerable areas.

All operators must have a valid permit in order to mine. Very detailed information must be submitted on the land and ecology, the operator's legal status, financial situation, and past history of complying with the law and the proposed mining and reclamation operations. To obtain a permit, an operator must meet all the requirements of the SMCRA and reclaim the land in compliance with the standards of the Act and implementing regulations. Under the provisions of SMCRA, the Office of Surface Mining has the responsibility of establishing a federal bonding system, and of evaluating and monitoring the bonding systems implemented by each State. The bond is a sum of money deposited as a guarantee against the failure to cure environmental damage resulting from mining.

S509 of SMCRA contains specific provisions regarding the amount and form of the bond, and the extent of liability. OSM's policy under recent Administrations has, however, been to allow the States maximum flexibility and has not discouraged the use of alternative bonding systems.

Subsequent to the Act, the OSM has introduced regulations to make it easier for operators to obtain bonds. First, OSM has introduced phase bonding as a possible solution to the objections from the surety industry to the long bonding periods. Under the phase system a regulatory authority may approve bonds from different sources covering the various phases of reclamation work. SMCRA provides for the release of performance bonds in three phases according to the following schedule.

- Phase I, 60% of the bond may be released after backfilling, regrading and drainage control is completed.

- Phase II, additional bond may be released after other reclamation activities are complete including establishment of vegetation (provided a sufficient amount is withheld to allow re-establishment of vegetation by a third party up to 5 to 10 years later).

- Phase III, the bond is fully released upon completion of all reclamation work and after all operator responsibility has expired.

The bond amount is determined by the regulatory authority and must according to the statute be based on the following factors:

- reclamation requirements of the approved permit;

- probable difficulty of reclamation, including the consideration of topography, geology, hydrology and revegetation potential;

- amount necessary to assure completion of the reclamation project in the event of forfeiture.

Cost estimates must include projected increases in order to reflect the real cost in years hence.

The regulations also allow the regulatory agency to adjust the amount of the bond as the area covered or costs of reclamation change.

The Act provides only that the liability exists for the duration of the operation and until the operator's responsibility for establishing revegetation has expired (usually 5 years after seeding but can extend for 10 years).

A common State variation to the Federal scheme sets a fixed per acre bond amount and establishes a special reclamation fund, supported by a levy on the output of all coal production in the State as a backup source of reclamation funds. Thus, in the event of an operator's failure, the bond would be forfeited and any costs of reclamation beyond those covered by the forfeited bond would be paid from the State fund.

SMCRA requires at least one complete inspection per quarter and one partial inspection per month without advance notice to the operator. SMCRA also provides for special inspections where citizens complain about hazards or violations at a particular mine. The inspectors responsible for conducting these inspections are vested with summary cessation powers. If an inspector finds an imminent danger to public health or safety or imminent environmental harm, the inspector must issue a cessation order against the particular part of the mine. In the case of a violation, the inspector must issue a Notice of Violation allowing no more than 90 days to abate. If the violation is not abated within the time set, the inspector must issue a cessation order and impose whatever obligations are necessary to cure the violation. Operators are also liable for stiff penalties. An operator who engages in repeated, negligent or wilful violations, may have their permit suspended. Intentional violators may be held criminally liable.

An important provision of the Act is the Abandoned Mine Reclamation Fund which is administered by the Secretary of the Interior. All coal mining operations subject to the Act are required to pay a levy per tonne of coal produced. S.402 of the Act requires that 50% of the reclamation fee collected must be allocated to the State from which it was collected if there is an approved reclamation programme within that State. The Secretary of State also has the discretion to allow all or part of the 50% allocation to be used by the State for the construction of public facilities in communities impacted by coal development and where extraction has ceased.

Canada : Province of Ontario

In Canada, legislation in relation to minerals restoration is formulated at Province level.

Southern Ontario is the highest demand area for mineral aggregates in Canada with Metropolitan Toronto and its surrounding areas consuming over 55 million tonnes per annum. There is pressure of demand for aggregates as a consequence of demographic and economic growth in this area and environmental resistance to quarrying and sand and gravel operations, somewhat along the lines of that experienced in London and the South-East of England and in the regions of Britain's other major conurbations.

The principal legislation controlling aggregates extraction was the Pits and Quarries Control Act 1971 which has been replaced by the Aggregates Resources Act, 1989. A licence from the Provincial Government is required to extract minerals from individual sites (this is additional to planning decisions, which are dealt with at the local level) and provisions are incorporated within the Acts with respect to rehabilitation of sites. These provisions include the payment of financial securities by the operators to the Provincial Government.

Under the Act, a licence application must be accompanied by a site plan which includes, amongst other things, progressive and ultimate development and rehabilitation. Information is also required for the ongoing operations of mineral extraction. Research has indicated that since 1971 the amount of rehabilitation has increased substantially and that the quality of this rehabilitation is also improving in a progressive manner. It has been concluded that between 60 and 70% of all agriculture rehabilitation has been successful, based on pre and post-extracted soil capability. Basic standards of rehabilitation are set down by the Provincial Government and operators must follow these.

New provisions in the 1989 Act require an increase in the quality and quantity of rehabilitation and improved powers to suspend a licence in the event of default. Each site is inspected at least once a year and this may increase at times of particular activity. The 1989 Act also introduces the idea of rehabilitation of abandoned sites with the funding of these rehabilitation schemes for abandoned sites coming from the licence fee payments of current operators. In making a decision on whether to issue a licence, the Minister of the Province must have regard, inter alia, to the suitability of the progressive rehabilitation and final rehabilitation plans for the site. Terms and conditions may be attached to the issuance of a licence.

Regulations under the 1971 Act required that a security (set at 8 cents per tonne since 1982) of the material removed per year from the property was deposited with the Treasury of the Province. Payments had to continue until such times as the licensee had a minimum of $1000 per hectare of land requiring rehabilitation on deposit with the Treasury. Any surplus over and above the minimum amount required may then be claimed by the operator for expenses incurred in the progressive rehabilitation of the site. In the event that rehabilitation did not occur, payment had to continue to a maximum of $6000 per hectare requiring rehabilitation. In this way, it was in operator's interests to undertake progressive rehabilitation to reduce the size of the security deposit. Once final rehabilitation was completed, any security remaining on deposit was refunded to the operator. In the event that rehabilitation was not undertaken or not completed to the Minister's satisfaction, the operator forfeited the money remaining in his account.

The money held in the account is held against the licensee's name and under the 1989 Act it earns interest. Also under the 1989 Act, if the amount in the account is not enough to cover the restoration costs then a debt is due to the Treasury which would be recoverable by the Province through the courts in the normal way. The size of the security payments may be varied from time to time.

The Province has a number of inspection officers and each licensed property is inspected regularly. These officers have substantial authority including powers to review records and lay charges.

In addition to the security payment under the licence, there is also the licence fee which under the new Act is set at 6 cents per tonne. Of this 6 cents, 4 cents goes to the local municipality in order to cover costs incurred locally as a result of aggregate operations. Additionally, ½ cent goes to the county or region and 1 cent to the Province. The remaining ½ cent goes to "an abandoned pit and quarry rehabilitation fund". This fund may be disbursed for purposes of conducting surveys or studies in relation to the rehabilitation of abandoned pits and quarries, and for actual rehabilitation of the abandoned sites.

An objection by anyone affected by the issue of licence requires the Minister to refer the matter to a quasi-judicial Tribunal. One drawback to the licencing process is the considerable length of time it takes for the necessary approvals, with two to five years and considerable sums of money being quite common.

Under the 1971 Act the Minister could suspend the licence only where the operation of the pit or quarry constituted an immediate threat to the interests of the public. Under the 1989 Act, the Minister is able to suspend the licence immediately for any period of time up to six months until a licensee has complied with a notice of violation of the Act or its regulations.

In 1991 Ontario introduced a revised Mining Act incorporating rehabilitation requirements covering metallic minerals which also has financial security provisions.

Australia

Regulation of minerals and mining activities is primarily the responsibility of state governments. As in the United States, the most developed systems of control relate to coal extraction. Coal is mined through both underground and surface methods in several Australian states. But two states, New South Wales and Queensland dominate production. The mines are both Government and privately owned. These two major coal producing states have adopted significantly different approaches to the regulation of surface coal mining.

In New South Wales regulation of the surface extraction of coal or the surface effects of underground mining has existed since 1973 with the passing of the Coal Mine Act. Under this Act the State Department of Mines leases coal to a private operator to extract and the leases impose environmental requirements including restoration.

The environmental performance standards in the lease in NSW are fairly standard and general, although discretion is left in the hands of the District Inspector. Operators may also be requested to produce an Environmental Impact Statement (EIS). The typical lease will include requirements for topsoil removal and storage, depths of replacement soils etc. The lease generally prohibits slopes on reclaimed land greater than 10 degrees and requires vegetation to be established as agreed by the Soil Conservation Service (SCS) and the operator. The operator must also submit an annual report setting out details of the reclamation work carried out in the previous 12 month period. In addition to ensuring compliance with the conditions of the lease, the operator must post a bond within 28 days of lease approval. Violations of lease conditions can lead to forfeiture of the bond. Formal enforcement action for breach of lease conditions is rare.

There has been some criticism of the rehabilitation programmes by environmental groups. The philosophy in New South Wales and the Department is that mines should be "rehabilitated" not restored to its previous shape or necessarily its previous use. Thus, the land is often reshaped to match the surrounding countryside rather than returned to the approximate original contour as required by United States law, for example.

The Queensland Government has been committed to the rapid expansion of coal production. To implement this policy, the Queensland Government has attempted to attract large multi-national energy companies to develop mines.

Little attention has, however, been given to restoration. Although some mines are controlled by franchise agreements between the companies and the State the agreement will contain very few restoration conditions. Once negotiated these franchise agreements are enacted as State law and override all other laws that would apply to mining operations, including any potentially troubling environmental requirements.

If no franchise agreement is involved, mining in Queensland is regulated under the Mining Acts of 1968 and 1971 which include some provisions for backfilling the pits and revegetation. However, because of the low population density and the lack of competing land uses, revegetation, soil productivity, and other environmental consequences of the mining activity do not receive high priority. Both formal enforcement action and citizen participation in the regulatory process are rare. Inspections are conducted by inspectors from the Department of

Mines who work out any problems with the companies informally. There is very little citizen involvement in the regulatory process. The few organised environment groups have focused their attention on more pressing environmental issues. Consequently, there is little attention paid to the environmental effects of surface mining for coal in Queensland.

3.7 Principal Findings from the Experience with Financial Guarantee Schemes

None of the examples considered provides direct experience that can be used to assess how a comprehensive scheme might work for guaranteeing the costs of mineral restoration in Great Britain, but the various examples discussed in this chapter do permit insights and general lessons to be learned.

There are two basic approaches to guarantees: decentralised, where a variety of individual guarantees are given by operators to the regulating authorities and which may be called in specified circumstances; and, centralised where some higher level of authority or agency, which may be Government, also in specified circumstances assumes the responsibilities for restoration.

For decentralised schemes, there appear to be no fundamental problems in establishing markets in guarantees. Experience suggests that smaller operators do have more difficulty in obtaining cover, or are obliged to bear higher costs since they are perceived to have higher risks of default. Where guarantees are issued by third party institutions there must be clear criteria for defining a default and release of the guarantee, else the guarantee becomes worthless.

One such criterion is financial failure and all guarantee mechanisms looked at in this chapter accept this as a valid case of default.

Technical failure presents rather greater problems of definition of default. The experience under voluntary Agreements and the South Wales Local Acts does not offer a basis for assessing whether the bonds and other financial guarantees do provide cover for technical failure since no bonds have been called for technical reasons. Contract experience from the building and construction industries suggests that the restoration scheme needs to be defined in very great detail before failure can be demonstrated.

A similar situation exists with the restoration bonds experience from the United States. In the US the standards of restoration have been specified in detail, and an expensive system of inspections is in place to ensure that works carried out by operators satisfy the inspectors and meet the specified standards.

There is limited experience of centralised systems of financial guarantee. Voluntary arrangements where operators come together as a group to organise a scheme such as provided by the example of the SAGA mutual fund, are likely to be problematic as a basis for moving forward for a comprehensive scheme since groups of operators will have an incentive to exclude from their group the operators with a high risk of default. The SAGA scheme works well for its members, but such schemes are unlikely ever to be comprehensive in their coverage.

Where Government establishes a scheme, the experience of the Ironstone Restoration Fund suggests difficulties in setting the fees or levies for the Fund. Government may not be able to move quickly enough in the face of changing costs to cover its liabilities. Additionally, a centralised scheme which takes over the liabilities for restoration, whether in whole or in part, removes restoration responsibilities from the operators and hence risks creeping costs.

A distinction may be drawn between centralised schemes which are intended to take over responsibility for restoration as in the Ironstone Fund case, and centralised schemes which are responsible only for the costs of defaulting operators. The distinction is essentially between a Fund for restoration and a Fund for insurance against default. There were no examples of an insurance fund.

4 ALTERNATIVE FINANCIAL GUARANTEE MECHANISMS

This chapter reviews options for financial guarantee mechanisms and evaluates them against restoration objectives. It starts with an assessment of the current magnitude of restoration liabilities in order to provide the context for the assessment of the costs of alternative mechanisms.

4.1 Restoration Liabilities

In order to establish the broad order of restoration liabilities, estimates have been assembled of total costs by mineral type.

The total restoration liability is the product of estimated number of sites ceasing operations and the average cost of restoration per site. Sources of evidence used in the estimates were as follows:

- Number of sites ceasing operation: data from the postal survey of local authorities, covering the period 1982-90.

- Restoration cost per hectare: from discussions with industry and ADAS. Costs are highly site-specific and depend on technologies used. The estimates are generalised but are consistent with a high standard of restoration and the requirements of the 1981 Minerals Act and advice in MPG7 and vary with afteruse as follows:

	Average cost
Agriculture	£20,000/ha
Amenity	£15,000/ha
Forestry	£16,000/ha
Other uses	£15,000/ha

Note that these cost figures may not apply in particular restoration situations, nor to problems encountered with certain mineral sites, for example where tipping has been substantial.

- Mix of afteruses by mineral type: based on the DoE 1988 Survey of Land for Mineral Workings. These data are for England, but are assumed to apply to GB.

- Average size of site: also based on the DoE 1988 Survey.

Table 4.1 indicates that assuming all sites ceasing operations are restored, total annual costs are about £150 million. Nearly 80% of this liability is for sand and gravel and opencast coal sites.

4.2 Purpose of the Financial Guarantee

The purpose of the financial guarantee is to act as a safeguard for the public that restoration will take place and will be of a satisfactory standard. In the event that an operator defaults on conditions for restoration or aftercare of a site, the financial guarantee ensures resources are accessible to cover the expenditure necessary to implement the conditions. A guarantee "scheme" could assume all the liabilities for restoration and hence would cost about £150 million p.a. Or it could assume the risks of default only, in which case the costs of the scheme would be a lesser amount.

Table 4.1 : Annual restoration costs by main mineral type (Great Britain)

	Estimated number of sites ceasing p.a.	Average site size ha	Average costs per ha £000s	Average cost per site £000	Total industry liability p.a. £000s
Chalk	11	16.9	17.6	297	3271
China clay	1	44.9	16.0	720	720
Clay/shale	20	20.3	17.6	357	7149
Coal (opencast)	63	51.7	19.2	993	62538
Gypsum	1	36.8	19.2	708	708
Igneous rock	4	17.8	15.5	276	1103
Limestone	17	22.1	18.7	414	7032
Sand and gravel	144	21.7	17.9	388	55808
Sandstone	7	8.4	18.0	151	1056
Slate	1	8.6	15.0	129	129
Vein minerals	23	12.4	19.0	235	5406
Others	11	18.1	19.5	354	3891
Weighted Average/Totals	303	26.6	18.3	491	148773

Note : Numbers may not total due to rounding

Default could be as a result of financial failure or similar reasons, or as a consequence of technical failure such as where the operator had undertaken a poor job, used inappropriate techniques, devoted insufficient resources for the task, had not complied with the agreed scheme etc. It is for consideration whether both kinds of default, financial and technical, should be incorporated within a scheme to safeguard restoration or whether only financial failure should be guaranteed.

Evaluation Criteria

The basic requirement is for a guarantee scheme which satisfies the planning authority requirements on security with respect to resources and gives appropriate access to funds, but only when default occurs. The arrangement or scheme should not impose excessive costs on operators, should work fairly in the sense of equal treatment of operators by different planning authorities, and should address problems of default on conditions already imposed and not some other problem (such as achieving restoration to new standards, unless these are imposed in the normal course of modernisation).

Basic Options

Based on the review in Chapter 3, alternative instruments for financial guarantees may be categorised in two broad ways:

- **Decentralised** - involving individually determined and agreed instruments where the operator provides the planning authority with an appropriate and sufficient security at the time of granting of planning permission, and where the instrument of the security can take one of a number of different forms. This would be similar, for example, to the approach adopted under existing voluntary Agreements and under the South Wales Local Acts. The operator is responsible for restoration and the security would be called only in the event of a recognised default by the operator.

- **National** - where a comprehensive, industry-wide fund or funding mechanism is established, probably but not necessarily on the initiative of Government, which provides the security for restoration costs against default by operators and for which there are two principal variants: restoration funds such as the Ironstone Restoration Fund, and insurance funds, of which an example is the SAGA mutual fund, but each of which has a number of different forms.

4.3 Decentralised Bond and Guarantee Instruments

Planning legislation would be needed to grant powers to permit planning authorities to impose as a condition of planning permission that a financial guarantee is lodged with the planning authority which will cover the risk of default on restoration and aftercare due to financial failure or other specified reasons. As with the South Wales Local Acts, flexibility would be maximised and costs reduced by leaving open the nature of the financial instruments that operators could place before authorities, there being no particular advantage in requiring all operators to provide one type of instrument. Safeguards would be necessary, for example, to ensure that authorities are not unreasonable in the types of financial guarantee that they would be willing to accept, and do not have unconstrained access to operators' securities. Set out below are various instruments that could be used with some of the situations in which they are likely to be practicable.

Bonds

The most commonly referred to instruments are bonds, analogous to performance bonds which were described in Chapter 3. There it was explained that there are, in fact, two types of bond:

- **On-demand bonds** - most commonly offered by banks. The bondsman simply agrees to pay the bond on demand and without proof of breach of contract or condition or any other factor which may be relevant to the two other principal parties. Because it is prepared to issue the bond on demand, the bank naturally secures itself for the bond sum and these bonds are virtually risk-free to the bank. Operators normally would be required to provide full security, in which case there is little advantage to them in this kind of bond. Very secure companies may be given a bond which "scores" against their borrowing facility in lieu of offering security.

- **Conditional bonds** - these are not secured except by a counter-indemnity usually in the form of a company written guarantee (which may be worthless in the event of bankruptcy or financial failure). The bondsman avoids payment of any claims until such times as he is fully satisfied that the terms of the bond have been met and where the beneficiary is able to prove his loss. The beneficiary (ie. the planning authority) will have to wait to get the money until the loss is proven. Because risk is involved, such bonds are available only to companies whose risks of default or failure are low. Both banks and the insurance surety market issues these bonds, but there are relatively few insurance companies and underwriters operating in this market. Clear-cut mechanisms are required to test for default, otherwise the conditional bond is worthless since the bondsman will resist payment for the duration of any doubt about default.

It may be possible to have a variant on the conditional bond, if that seems appropriate, where an independent expert certifies that the operator is in default. The bondsman would pay up to the planning authority as beneficiary only on the issue of the appropriate arbitration certificate.

The two principal types of bonds have somewhat different Planning procedure implications in the case of restoration. The "on-demand" bond gives to the authority the apparent advantage that, as bond holder, the money can be secured at any time. This will favour authorities greatly at operators' expense and, indeed, the only real constraint on local authorities calling the bond would be the prospect of a claim made by the operator in the courts against the authority that had acted unreasonably in calling the bond and for which compensation would be claimed by the operator. This might be an effective restraint on an authority holding an on-demand bond, since the fear of a compensation claim is often said to put off planning authorities from issuing stop notices for breaches of planning conditions, and the two situations are analogous. But against this there would be a real danger that there would be other authorities that would choose to call bonds in preference to pursuing operators through the normal enforcement channels. Without due safeguards, therefore, the on-demand bonds are potentially liable to supersede the existing planning enforcement procedures altogether.

Indeed, the prospect that existing procedures would be superseded applies not only to on-demand bonds but to on-demand securities of all kinds, such as cash deposits, personal securities etc. (see below), where the planning authorities have direct access to the security.

This fear that the on-demand bond might supersede enforcement procedures does not apply to conditional bonds where to call the bond the beneficiary proves his loss. In such cases, the bondsman would only be prepared to pay up when the terms of the bond have been met. Financial institutions will not be willing to involve themselves in the details of planning conditions and compliance assessments but will seek some form of procedural criterion for judging whether to permit the bond to be called. The most obvious and likely requirement will be that enforcement procedures had been exhausted, in which case there is no value in the bond since if enforcement takes place, the conditions to the planning permission will have been implemented without the bond. A variant would be the arbitration certificate mentioned above. The conditional or on-demand bond could be called where it was clear that the operator had failed financially, since this would constitute default. Indeed, it seems likely in practice that financial failure would be the only acceptable grounds for releasing the bond, as is the case for most building and construction performance bonds (see Chapter 3).

In effect, in the case of the conditional bond the bondsman is insuring against the insolvency of the company. He will only do so for companies having a low probability of insolvency. Even for companies that qualify for a conditional bond financial circumstances may quickly change and a six months credit risk may be considered normal, and three years is unlikely to be exceeded.

These timeframes are, of course, far too short in relation to the requirements of security for restoration. One way round these timing difficulties may be to have some kind of "evergreen" system, whereby the bonding is renewed on an annual basis but extends always for a further one or two years. Thus, in the event of a bond not being renewed the planning authority would still have a bond which is valid for the further period during which time alternative arrangements would need to be sought, or the bond could be called. Evergreen bonds would be possible and feasible to introduce, but would add to the cost of the bond.

Fees for bonds will depend on circumstances. Contract performance bonds cost about 1-3% of value p.a., and similar fees have been charged for restoration bonds in South Wales, although this may be a poor guide as to the fees that would be charged for all restoration bonds.

The market in bonds generally favours the larger companies, and those with greater financial security and will inevitably discriminate against the smaller operators who will be perceived as a higher risk of insolvency and who may not be able to obtain bonds at all. Thus, introduction of a system of compulsory bonding would lead to reduced competition amongst mineral operators by forcing some smaller companies out of business if they could not raise bonds or when faced with higher costs, and certainly would be a barrier to the entry of new firms into the mineral extraction business unless backed by the resources of larger firms. This is a clear and convincing reason for not restricting the type of instrument of security to planning authorities specifically to bonds.

A crude calculation of the possible costs to industry of such bonds is presented in Table 4.2. assuming that bonds could be obtained and were held on all sites and that all sites were due to be restored on completion of working.

These and other calculations of costs do not include estimates for restoration of china clay and deep-mined coal tips.

Overall restoration liability is the product of average restoration costs and the total number of sites as derived from the DoE 1988 Survey (England) and 1988 Directory of Mines and Quarries and British Coal Opencast Executive (Wales and Scotland). This suggests that total eventual restoration liability for all mineral types is approximately £2.1bn, with sand and gravel, limestone, opencast coal and clay/shale facing the largest restoration liabilities.

Thus, the estimated cost in bond fees to the minerals industry would be in the range £21-63 million per year, assuming an annual bond fee rate of between 1% and 3%. This is a "deadweight" loss to the industry, to be incurred in addition to the £150 million p.a. costs of restoration and for which the principal benefit to the local authorities would be safeguarding against financial failure.

Operators unable or unwilling to organise a bond should be permitted under a decentralised scheme to offer local authorities security in another form. There are a number of options, as follows.

Cash Deposits

The operator could place a cash deposit with the planning authority. This could be a joint account with, as a safeguard, an agreed arrangement for access to the funds in the account, possibly including some form of discharge agreement or arbitration arrangement. One such account is known as an escrow account where both sides have access to the deposit subject to meeting certain requirements. The operator would expect to draw on the account to finance reclamation work.

This type of guarantee arrangement would involve, of course, the removal of capital from company use. The cost of such an arrangement to an operator would be the opportunity cost of the capital tied up in this way, equivalent to the lost interest or returns. The magnitude of industry costs would depend on who received the benefit of the interest on the cash put on deposit. This cost falls if the company remains the full beneficiary of the interest, although assuming overall lost return of 3%, the industry cost would be about £47 million p.a., which makes this form of security broadly as expensive as bonds.

Table 4.2 : Estimated costs of a bonding scheme for current mineral sites

Mineral type	Total number of sites[1]	Total restoration liability[2] £000s	Industry Cost of Bonding £000s per year	
			1%	3%
Chalk	216	64152	642	1926
China clay	53	38160	382	1146
Clay/shale	510	182070	1821	5463
Coal (opencast)	300	297900	2979	8937
Gypsum	22	15576	156	468
Igneous rock	224	61824	618	1854
Limestone	587	243018	2430	7290
Sand and gravel	1755	680940	6809	20427
Sandstone	380	57380	574	1722
Slate	43	16641	166	498
Vein minerals	124	29140	291	873
Others	361	127794	1278	3834
	4262	2092642	20926	62778

Note : (1) Derived from the DoE 1988 Survey of Land for Mineral Working for England and BGS Directory of Mines & Quarries for Wales and Scotland.
(2) Using data from Table 4.1 in relation to the costs per site.

Regular Payments

A variant on the cash deposit is for the operator to build up an individual restoration fund by payment of a levy based, for example, on the output of the site such that the fund will grow as the mineral is extracted and restoration costs increase. This would be somewhat along the lines of the Ontario scheme (Chapter 3). In its extreme form, each planning authority could establish its own restoration fund consisting of deposits by a number of operators in its area, either as a free-standing fund or as a series of individual funds, one for each site.

There may be tax advantages for smaller operators in such an arrangement if these payments could be regarded as restoration costs, and hence be tax deductible expenditure. Against this advantage, operators would loose the flexibility in the use of funds, and costs would be broadly similar to the case of cash deposits.

The disadvantage from the local authorities' standpoint is the risk of default in early years, as funds would be insufficient to meet restoration liabilities in full unless the site was being progressively restored.

Title Deeds Etc.

An alternative to providing a cash is security in the form of assets such as title deeds over property or liens on particular assets, which are then held by the planning authority or by a suitable third party. These forms of security are known to be common for existing planning Agreements and under the South Wales local acts. The costs of these securities is not directly observable.

One factor for the local authorities to consider will be the availability of assets of the company which are not already held as security against other liabilities. It may be difficult legally for the authority to tie up the assets specifically for purposes of restoration in the event of bankruptcy and consequential calls on the assets from other creditors.

Charges on company assets would require registration with the Registrar of Companies. Planning authorities would find it necessary to search for other charges on assets with the Registrar which might have a prior claim. It would be dangerous, of course, to link the security to the mineral assets in the ground. The tangible assets of operators might not be sufficient to cover the extensive costs of restoration, particularly in the later years of the extraction of material from a site when the site's economic worth has been removed.

Some small operators may need to give personal guarantees to planning authorities, which might involve the authorities in difficult decisions concerning calls on the personal wealth of these operators.

Parent Company Guarantees

For operators of sufficient credit-worthiness, and which have a parent company of high credit standing, a parent company guarantee may be both simple and cost effective to provide. However, evidence collected in the personal interviews suggests that parent companies are often unwilling to permit their balance sheets to bear the liability of guarantees for subsidiaries, and even the most credit-worthy major company would find difficulty in offering a guarantee for a period of beyond about three years. Accordingly, such guarantees would need to be renewable, and in the event that the local authority was not satisfied at the point of renewal, some alternative form of guarantee would be required.

Mutual Funds

A further form of security and one which has features of both centralised and decentralised schemes would be a guarantee scheme covering several operators where risks are spread and the group offers the security. An example is the SAGA scheme. There may be problems with establishing such mutual funds in the earlier years, until such times as the fund is large enough to meet potential liabilities. There will be some difficulties also with this arrangement if the authorities do not accept the fund as adequate security. Indeed, some authorities have requested that SAGA members take out additional forms of security in voluntary Agreements which SAGA has resisted. There is little doubt that if the planning authorities do accept the mutual fund, it can be a very cheap method of providing security for groups of operators, particularly those with low risk of default.

Summary of Decentralised Forms of Guarantee

Leaving the decision to the individual planning authority and operator as to which form of guarantee will satisfy the requirements of the planning condition has the advantage of giving maximum flexibility to the way the system would work. Indeed, it would be unreasonable to impose a requirement that any particular kind of guarantee should be taken out under the decentralised scheme.

On the other hand, flexibility would offer maximum scope for variation in standards and approach to be adopted by the authorities. Giving local authorities the power to impose financial guarantee conditions could, unless checked, lead to some operators failing to proceed with extraction on sites in one authority's area, which would be available in an adjacent area for no other reason than a different perspective was taken on the appropriateness of particular forms of guarantee.

Of course, as a safeguard, an operator could appeal against unreasonable financial guarantee conditions being imposed but this hardly meets the point that different standards will be applied by different authorities.

Authorities would need to have in place mechanisms for assessing different operators' financial status in relation to the mineral extraction and restoration tasks being considered in the planning application. Additionally, there would need to be costs of administering the process of guarantee issue which, in the case of some authorities with extensive mineral operations, would not be trivial. There may, therefore, be public expenditure implications.

It will be an important requirement of any decentralised scheme that adequate safeguards exist and in particular that authorities do not have unconstrained access to the security and hence to restoration funds.

The costs of some decentralised systems of financial guarantee are high. All operators need to be covered by the arrangements since any one of them may default and until the market has established the patterns of risk there will be considerable deadweight of costs. In extreme form the operator may need to set aside cash, or its equivalent, equal in value to the restoration costs and for the duration of development. The costs of this would be at least equal to the foregone interest earned. Even in instances where a guarantee can be purchased from a financial institution, the institution is likely to adopt a conservative attitude to the risks involved when considering operators' requests on an individual basis.

The costs involved to the operators, and particularly smaller ones, to provide the guarantees are such that the pursuit of a financial guarantee system on a decentralised basis needs to be carefully weighed in the balance against the benefits to be realised.

4.4 National Scheme : Restoration Fund

A national restoration fund would provide a full guarantee for restoration, since the fund would assume all responsibility for restoration. In order to provide full security, the fund itself would need to be underwritten in some way. A voluntary restoration fund covering the whole of the minerals extraction industry is unlikely to be established because of the difficulties associated with bringing all operators within the umbrella of one voluntary organisation, although it is just possible that the CBI Minerals Committee would provide a starting point for a voluntary scheme as all trade associations are represented.

More likely, a restoration fund approach would have some kind of Government backing, being initiated by Government and, in the early years at any rate, underwritten by Government. The fund need not be a permanent call on public expenditure if it is operated according to the principle that, taking one year with another, income from fees or levies imposed on operators shall exceed expenditure on restoration and administration. Unfortunately, this was not the experience of the Ironstone Restoration Fund, which charged too little.

The restoration fund would need to have all operators as members. A pre-requisite for mineral extraction would be membership of the fund. This would be equivalent to the issue of a licence to extract minerals by the fund management, with no operator permitted to undertake extraction activities without having the licence and being a paid up fund member.

A trade association fund would be unacceptable, since there would be a presumption by existing members to exclude those that were not members, other than those with low risks where the contributions to the fund would exceed the expected payments from the fund. The very purpose of the fund would be to cover restoration costs (or a topping-up element of them) of all extraction activities and all operators need to be included. Of course, a Government restoration fund need not accept just anyone as a member. But qualification criteria would need to be carefully defined to draw a balance between restraint of trade arising from a barrier to entry into the industry (or exclusion from it) and risk of restoration defaults.

If the Ironstone Restoration Fund is any guide, this kind of arrangement is risky. That fund failed largely because the income was insufficient, at the end of the day, to bear its share of costs of restoration of ironstone sites when ironstone became no longer economic to extract. Failure arose from the inability of Government to set a levy at a level sufficient to cover costs.

Whilst separate funds could be run for individual mineral types, probably a composite fund embracing all minerals would be best since the risks are more widely spread across different industries. However, the levy to be imposed on mineral operators would need to be subject to variation by type of mineral. A levy based on output, whether measured by tonnage, volume or even value in relation to restoration costs would vary significantly from one mineral type to another.

The comprehensive restoration fund concept suffers from the disadvantage that there is insufficient restraint on restoration costs to creep upwards. Responsibility for restoration is taken away from the individual operators, and fund managers are unlikely to be able to restrain costs to the same level as would be the case with individual operators seeking the most cost effective solutions.

Another objection to the restoration fund approach is the size of the flow of funds that would be involved, up to about £150 million p.a. with commensurate administration costs to care for them.

Stevens argued strongly against restoration funds, largely on the grounds that they involve the fund taking over responsibilities of others unnecessarily, and for which there would be an added administrative burden. Stevens was of the opinion that restoration funds were irrelevant to the concept of financial guarantee intended to deal with default. Stevens recommended against a general fund set up along the lines of the Ironstone Fund.

4.5 National Scheme : Insurance Fund

A second national approach to financial security for mineral restoration would be a fund which would assume the risks of individual mineral operators defaulting through financial failure and other specified conditions. The fund would meet the restoration costs when default occurred. It is difficult to imagine how a comprehensive scheme along these lines could be established voluntarily since, as with the restoration fund, a voluntary arrangement would have an in-built incentive to exclude some operators.

The insurance fund approach, covering the risks of default, would have many characteristics in common with the restoration fund in terms of criteria for membership and attendant licensing arrangements.

The commitment by Government, in terms of underwriting, would conceivably last for the first few years only until the fund is fully established. The financial flows would be quite modest since only costs of default are to be covered. After a short time the fund could be fully floated off when it becomes self-financing.

In the absence of sufficient knowledge to the contrary about individual operators' risks of default, the financing of the scheme would involve a "rough justice" fee or insurance levy imposed on all operators. The low risk operators, principally those which are larger and more dominant in the market, would cross-subsidise the high risk operators, of which many would be smaller companies.

In all probability after a few years, sufficient evidence will be built up on the nature of risks so that the charging system could be discriminatory, taking into account the probability of individual default and the default costs. But in the early years the charges would probably need to be common to all members of the fund with payment amounts based on a formula, of which a link to the operators' output would be simplest. Output could be measured by tonnage, volume or value. Other possibilities for setting the levy include linking fees to area worked and rateable value. This formula approach would be necessary if the fund administration is to be kept simple.

The general principle behind the Insurance Fund approach is that it is cheaper to underwrite a "book" or large number of risks taken together than to do so for a large number of risks taken individually. This is because on an individual basis each risk has to be evaluated for which there is an administrative or processing cost, and a pessimistic or "risk adverse" view likely to be adopted by the individual insurer. When insuring a book, however, the justifiable pessimism in relation to a few individual operators or circumstances will be counter-balanced by the many successes of the others.

The Ontario security deposit system is a variant of the Insurance Fund approaches. All operators pay into the fund, but in Ontario their accounts are kept separate. Each account is built up as levy payments go into it. Operators can call on their account to finance restoration costs when they occur. However, there is also a balance in the account (until such times as the restoration has been completed) in order to ensure that residual costs of restoration can be covered.

What an insurance fund would need to do, however, is widen the scope of this Ontario arrangement to ensure that some amount of each operator's account can be transferred between accounts by the fund managers to cover any deficits that may arise in defaulting operators accounts.

A difficulty with the insurance fund approach is that wilful operators have no incentive to continue with their restoration, and defaults are bound to increase unless safeguards are built into the system. To some extent the Ontario scheme solves this problem since the operator forfeits his contribution to the fund if he fails to proceed with the restoration.

A variant on the insurance fund approach would be a **National Bonding Institution**, which would offer operators bonds which could then be taken to the planning authorities as evidence of security. The differences between the insurance fund and the bonding institution is that the latter would require a counter-indemnity of some kind before issuing the bond. Operators would not need to be members of any particular organisation or scheme but simply would seek their financial security as the need arose in much the same way as they may seek it in the banking or insurance market. The security would be standardised and, of course, compulsory under this scheme.

The bonding institution would be more complex to operate than the insurance fund, would certainly require greater administration and overheads, and would raise objections from those parts of the mineral extraction industry who could secure bonds in the private market without difficulty (ie. the larger operators) and who would be obliged to incur charges that in all probability would be higher than they could obtain elsewhere.

Stevens regarded the insurance fund approach as potentially offering advantages over individually negotiated bonding arrangements in the open market.

Costs of a National Insurance Scheme

In order to assess the cost implications of a national insurance fund scheme for operators, a financial assessment has been made which takes into account the potential calls on a fund and the levy that would need to be charged to operators to ensure that, taking one year with another, the fund breaks even.

The component costs of an insurance fund are:

- payments for the costs of restoration on sites where there has been a default;
- interest charges associated with having a sufficient capital reserve on call to meet unexpected eventualities; and

- administration.

The costs of a capital reserve are assumed to be the interest charges for borrowing up to one-half of a year's worth of expenditure requirements. Administration will have a fixed cost component and a component which is variable, dependent on the level of insurance activity. Cost estimates have been made based on potential administrative charges of a fixed cost of £400,000 p.a. (equivalent to a core staff of about eight, including all overheads) and a variable cost equal to 15% of restoration expenditure.

In the calculations two basic assumptions have been made concerning the nature of the fund. The first is that the principal objective is to cover the risks of operators' financial failure. The second assumption is that, in addition to financial failure, the fund will finance technical failures of operators to restore to satisfactory standards. Survey findings reported in Chapter 2 indicate that financial default rates are small relative to the technical failure rates, and broadly are of the ratio 1:7.5 (ie. for each financial default there are about 7.5 technical defaults). Default rates by mineral type have been used in assessing costs, based on the survey results.

The results of the calculations are shown in Table 4.3. They are expressed as a levy per tonne of output of each mineral type (derived from BSO Statistics).

If financial failure alone is the risk to be covered by the insurance fund, the size of the fund would be about £2.8 million per annum and administration would take up about 25% of total expenditure. The levies charged for individual mineral types would vary significantly, depending on the restoration cost per site and the probability of failure. However, the levies would be small in relation to typical royalty payments for mineral extraction, for example. They are also small in comparison with the costs for individual bonds, as assessed in Table 4.2.

However, if both technical and financial failure were to be covered by the insurance fund, the size of the fund would rise to about £18.6 million per annum and the levies that would need to be charged would rise accordingly, although would still be only about 4p per tonne overall, again with substantial variations by mineral type.

Table 4.3 :   Costs of a National Insurance Fund

	Purpose of Insurance	
	Financial Failure	Technical and Financial Failure
Annual Fund expenditure of which:	£2.8 m	£18.6 m
Administration	£0.7 m (25%)	£ 2.7 m (15%)
Interest charges	£0.1 m (3%)	£ 0.6 m (3%)
Restoration expenditure	£2.0 m (72%)	£15.3 m (82%)
Average annual levy:		
All minerals (av.)	0.64 p/t	4.22 p/t
Chalk	0.69 p/t	4.55 p/t
China Clay	0.49 p/t	3.25 p/t
Clay/Shale	1.18 p/t	7.79 p/t
Coal (opencast)	1.49 p/t	9.80 p/t
Gypsum	0.20 p/t	1.29 p/t
Igneous Rock	0.06 p/t	0.41 p/t
Limestone	0.17 p/t	1.15 p/t
Sand and Gravel	1.22 p/t	8.05 p/t
Sandstone	0.20 p/t	1.29 p/t
Slate	0.99 p/t	6.54 p/t
Vein Minerals	204.95 p/t	1348.19 p/t
Others	6.56 p/t	43.16 p/t

Note :   Assumes number of sites ceasing as Table 4.1
Assumes Fund income and expenditure balance, one year taken with another.

The cost calculations reflect past rates of financial and technical failure, specifically those that were experienced in the 1980s. If an insurance fund were introduced, the rates of failure would almost certainly increase, reflecting the shifting liabilities for restoration costs, unless some penalty mechanism could be introduced to deter such a trend.

4.6   Conclusions on Alternative Mechanisms

Two basic approaches have been identified. A decentralised scheme would require operators to satisfy the requirements of the local authorities by lodging an individual financial guarantee. The alternative of a centralised or national approach would require that operators become members of a fund or scheme, for which they would pay a levy or fee and for which membership would confer benefits of security for the local authorities in the event of operator default.

Under a decentralised scheme there are many possible instruments of financial guarantee and it seems desirable to keep these options open and to let operators decide on which instrument is most cost effective for them, subject to satisfying the requirements of the local authorities.

Costs of a decentralised scheme could be high for smaller operators who face difficulties in obtaining financial guarantees or in raising securities. There may be some issues of restraint of trade to be considered in the overall evaluation of such a scheme.

Two centralised or national schemes have been considered. Restoration funds take responsibility away from operators and introduce an administrative charge and there are powerful arguments against a restoration fund.

A centralised insurance fund type of scheme has advantages and would appear to be substantial cheaper for the industry overall than a decentralised scheme based on individually negotiated instruments. A scheme could be designed to insure against operator financial failure, in which case the flow of funds would be very modest. If the insurance fund scheme were available for technical as well as financial failure the flow of funds and hence, operator charges would be substantially greater, and there would be a danger of rising costs and increasing operator default. Criteria for determining technical failure would need to be established.

The total costs to the industry of a decentralised bonding scheme would far exceed the costs of a national scheme. Estimates suggest about £3-19 million per annum for a national scheme compared with £21-63 million per annum for a decentralised scheme.

The need for a scheme, and the type of scheme that would seem most appropriate for the British Planning system, if one were to be introduced, are discussed in the next chapter.

5 CONCLUSIONS AND RECOMMENDATIONS

5.1 Limitations of Existing Procedures

The surveys which were undertaken as part of this study provide encouraging evidence that restoration performance under existing procedures is generally satisfactory, and appears to be improving. Nevertheless, about 27% of mineral sites ceasing operations are not restored to a standard which local authorities find satisfactory, or are not restored at all. This proportion remains broadly unchanged since the time of the Stevens survey although it seems likely that standards being applied to the assessment of what is satisfactory have increased.

However, the mix of reasons associated with failure to restore to a satisfactory standard has changed with a far smaller proportion of sites failing due to the implementation of the planning system than at the time of Stevens, and a far greater proportion failing due to diverse technical problems on-site. Financial failure of operators was small at the time of Stevens, and remains so today. In the current survey only 5% of sites where restoration was not to a satisfactory standard could be attributed to the financial failure of the operator.

Operators included in the surveys had different opinions and experiences to the local authorities. In the view of the operators, the incidence of failure to restore to satisfactory standards was far less than the incidence quoted by authorities.

Failure of Implementation of the Planning System

Limitations in the implementation of the planning system constitute about 35% of sites where there was a failure to restore to satisfactory standards. A variety of reasons were attributed to this. Some were the absence or inadequacy of conditions of planning permissions, where those permissions were historical. These reasons for failure could be reduced by implementation of the modernisation provisions of the 1981 Minerals Act, but authorities generally have failed to implement these provisions of the Act to a sufficient extent due mainly, it was stated, to inadequate resources for the task.

The other reasons for failure of the planning system were attributed to failure to monitor operators' activities properly, and failure to enforce operators to comply with planning conditions. Both of these factors may be attributed in part to problems of resourcing within the local authority planning departments although, additionally, the time and uncertainty associated with enforcement procedures have played a part.

The 1991 Act has enhanced the powers of local authorities generally as regards enforcement although it seems likely that these will not significantly reduce the complexity, delay or resource effort that is needed to undertake successful enforcement action.

Thus, the limitations of the planning system have less to do with the powers available to local authorities at the current time than their resources to fully implement the system.

Some special problems are not adequately dealt with by existing arrangements. The study identified borrow pits which cause local authorities particular difficulties as a consequence of their short-term nature and the lack of competence by the operators to restore sites. Vein mineral sites have caused pollution hazards arising from water outflows from disused mines, not commonly experienced with most other minerals but these are outside planning control. Private coal sites, in certain circumstances, also have presented difficulties because of the age of the sites and because often the mineral is being reworked from colliery tips with the effect that the value of the output is low relative to the magnitude of the restoration costs with a consequential tendency by operators to avoid undertaking restoration.

Although not a problem at the current time, some authorities express concern over restoration of deep mine British Coal sites in the event of privatisation given the absence of restoration conditions for many of these sites and the fact that there would be little incentive to acquire them should restoration conditions be imposed.

Financial Failure

Financial failure represents a small proportion of sites where there was failure to restore to satisfactory standards. Whilst in principle a planning authority may pursue the successor in title in the event of liquidation, this was generally regarded as being impracticable. The numbers in the survey were too small to ascertain whether there are significant associations between the incidence of financial failure and mineral type.

Diverse Technical Problems

Diverse technical and other problems linked to inadequate practices by mineral operators or their methods of working constitute the largest reason for failure to restore to satisfactory standards. In some instances, this arises from circumstances apparently beyond the operators' control. For example, some operators have taken over older sites where earlier working methods have made the task of restoration difficult and costly. In other cases, lack of expertise by the operator is the principal cause. Whilst no statistical association was found between failure to restore to satisfactory standards and the size of operators, authorities mostly regarded the lack of expertise as a problem of smaller operators. Technical problems mentioned encompass a wide range including poor drainage, uneven settlement, contamination and shortage of fill materials.

Problems which have occurred on some filled mineral sites due to settlement and contamination may lessen in future as a consequence of increased liabilities and stringency of licencing requirements under the provisions of the Environmental Protection Act, 1990.

Factors Leading to High Standards of Restoration

Operators interviewed in the surveys appreciated the need for restoration conditions and for restoration to take place to a high standard. They argued that most operators have extensive technical expertise of restoration, although many face an inheritance of sites which, by today's standards, have been poorly worked in the past and which makes restoration to contemporary standards difficult to achieve.

Current best practice is for planning conditions of the planning permission to be realistic and carefully worded and for the operator to produce a detailed plan for restoration and aftercare treatment linked to a financial and material budget.

The overall standard of restoration achieved will ultimately depend on the will of the operator linked to high technical standards and expertise. But the operator has to have in place effective planning, management and financial provision for the task and active liaison with the planning authority and relevant agencies. The local authority needs information from the operator on site investigations and the restoration and aftercare plan. The incentive to secure future permissions is a factor in operators' will to restore.

The preparation of environmental audits by operators should be encouraged as part of a commitment to raising environmental standards.

5.2 Scope for Improvement to Planning Practice

The study has highlighted limitations of current planning practice in the following areas:

- inadequate planning conditions for many historical sites;

- low level of enforcement action by local authorities against operators not in compliance with conditions;

- limitations of monitoring activity by local authorities, particularly at key stages in the workings of sites and during restoration.

Planning Conditions

Authorities have powers to review and modernise planning conditions under the 1981 Minerals Act. These are not being used enough. The conclusion of the survey was that the powers are sufficient but they give rise to unacceptable demands upon local authority resources, both manpower and financial resources in relation to potential compensation payments.

Evidence obtained in the operator surveys suggested that many operators are willing to accept some updating of conditions and that this occurs anyway as part of the process of obtaining new or revised permissions on sites. Indeed, many operators adopt a commitment to raising their environmental image by carrying out restoration beyond the terms of conditions on historical sites.

Nevertheless, a problem remains of the inadequacy of resources available to local authorities to update historical permissions. The approach adopted for IDO mineral permissions provides a model which could be considered by Government for extending the updating of conditions on some historical sites. Under such a scheme, operators could be given the opportunity to present a working and restoration plan to the planning authority on all sites where no restoration conditions currently exist.

Enforcement

The 1991 Act will improve enforcement procedures and powers. At this time, no serious deficiencies exist in relation to powers although the workings of the Act will require monitoring and assessment in due course.

The principal problems associated with enforcement arrangements during the period of the research, 1982-90, derive from the delay, costs and difficulties with procedures especially for those planning authorities which have undertaken enforcement action on an irregular and infrequent basis; the less enforcement action which is undertaken, the less well it is done.

Nevertheless, delays do occur, and have cost implications and can be frustrating and act as an deterrence to planning authorities from taking action. The system of enforcement does not cause failures to occur but clearly causes some problems of implementation of action against defaulters.

The principal solution is for those authorities which are failing to enforce adequately to consider upgrading the status of the activity and the grade of monitoring and enforcement staff. Consideration could be given to the setting up of specialist mineral enforcement responsibilities within authorities. These changes would require additional resources.

Monitoring and Management

Successful restoration and aftercare requires thorough site investigation and the preparation of an appropriate plan and, during the restoration process, proper monitoring and management.

For planning authorities to be effective, the operators must provide proper documentation. At the same time, authorities must engage in regular monitoring activity. Evidence collected in the surveys suggests that most authorities have well organised monitoring arrangements and have expertise available of the appropriate kind, but are unable to implement the monitoring at critical stages in the working of sites and in the implementation of the restoration schemes.

Again, to be effective, the authorities need to devote additional resources to the task within the framework of the powers already available to them.

Specific Problems

Specific problems have been identified in relation to:

- Borrow pits - where the principal need is to ensure that operators do comply to the same standards as for other minerals sites, and additional monitoring and enforcement effort by local authorities is required. Due to the short time-scales, contract performance bonding might be the best route to ensure compliance.

- Some coal sites, mainly those being reworked privately - where security against financial failure of the operators (due to the marginal economic nature of the activity) is mostly required.

Summary of Principal Planning Policy Issues

Existing planning powers available to local authorities are broadly sufficient to achieve modernisation of historical permissions, and the planning and implementation of appropriate restoration and aftercare schemes. Powers could be widened to embrace the introduction of conditions for restoration for sites which currently have none, somewhat along the lines of the IDO procedures, although compensation issues would need to be resolved.

The principal limitation at the current time on effective planning action is lack of resources available to local authorities to implement the system to its fullest extent. This is partly a matter of finance in relation to compensation provisions for modernisation, and the fear of compensation in the event of enforcement action which is successfully challenged. Additionally, however, there are problems of adequacy of staffing and manpower which are often insufficient to satisfy the demands being placed on them.

Problems do arise with financial failure of operators, although this is a small problem overall. Safeguards are required where there is an inability to pursue enforcement action against successors in title.

5.3 Policy on Financial Guarantees and Bonds

Existing Agreements

A few local authorities have adopted policies requiring financial guarantees and bonds from mineral operators. These policies are implemented through voluntary Agreements, and about 100-120 are estimated in existence in Great Britain, about two-thirds of which are in Scotland.

Although very modest in scope at present, the practice of trying to reach planning Agreements seems to be growing. There is no clear Central Government policy statement at present, although draft guidelines issued by the Scottish Office in July 1992 on land for mineral workings indicates that financial bonds are not favoured.

Whilst experience is limited there is a clear possibility that pressure for planning Agreements will grow. There are real doubts however that the Agreements achieve much for mineral planning authorities in practice. They appear to offer safeguards against operators' financial failure but are of doubtful worth in relation to technical failure to restore. There is a danger that they will be operated in an arbitrary fashion, not treating operators on a like-with-like basis. There is a cost both for the minerals industry and society more generally in the form of the reduced productivity of capital that is locked up in the guarantee mechanisms. Existing voluntary arrangements do not clearly evaluate the societal cost against the gains which are secured.

Accordingly, there is a case for rationalising the existing ad hoc arrangements involving Agreements, either by the introduction of a formal scheme for financial guarantees and bonds which would remove the need for Agreements altogether, or by clear guidance from Government as to the appropriate circumstances and conditions attaching to financial guarantee arrangements reached by planning Agreement.

If the Government wishes simply to clarify policy and to guide planning authorities' approaches to planning Agreements without changing the basic system, then this can be achieved by way of guidance notes or circulars. A new formal scheme would require legislative changes.

Basic Approaches to a Formal Scheme for Financial Guarantees or Bonds

The case for a new financial guarantee or bonding scheme can be made at one of two levels:

- either, as a replacement of existing enforcement procedures altogether, where the bond becomes the full security for compliance with conditions; or

- to supplement existing enforcement procedures where the bond is a limited safeguard against failure and, hence, becomes a component of the range of powers available to local authorities.

This distinction is of fundamental importance since it greatly governs the nature of the scheme, its objectives and costs.

Bonds as a Replacement of Enforcement Procedures

A comprehensive financial guarantee or bonding scheme which covers the risks of financial failure and which includes guarantees against technical failure would almost certainly become a replacement for existing enforcement procedures. Mineral planning authorities will not, in the main, choose to undertake lengthy, uncertain and potentially costly enforcement if they can have ready access to funds to undertake restoration and aftercare for any failure by the operator. It would be easier in most circumstances for a planning authority to call the guarantee or bond than to engage in enforcement action.

For a comprehensive scheme to work successfully there would need to be some radical changes to existing planning procedures. The financial guarantee or bond would be called when default occurs, ie. when the condition attaching to a planning permission was not met. The bondsman is not the best judge of these circumstances. Indeed, bondsmen would be unwilling to enter into the debate about whether restoration is or is not satisfactory, and would wish to rely on procedural criteria for release of the bond.

This presents no problem of substance as regards the financial failure of the operator. If the operator fails financially, then default has occurred. But technical failure requires the exercise of judgement, and this would be the cause of substantive changes to planning procedure. The issue is, who makes this judgement?

One option is to rely entirely on the authorities' view of the matter on whether restoration conditions have been met. In this case the financial security would need to be on-demand and would be called as and when the authority decides. It would be expensive for operators to obtain this cover and it is doubtful whether the resources of operators should be put entirely at the discretion of the authorities in this way.

If safeguards were to be introduced through an appeals procedure, the operator would need to argue against the unreasonableness of the authorities' decision to call the security; ie. that the conditions had, in fact, been complied with. In this event, the procedure would become similar in substantive respects to the existing enforcement procedures, and the value of the bond to the authorities would be modest, if not entirely removed. So, any system of appeals would need to be curtailed.

An option might be to have a technical arbitrator who would make the decision on the release of the bond. The arbitration procedure would require an independent expert, panel or institution to determine whether the conditions of the planning permission in relation to restoration and aftercare had been breached.

The US experience with bonding arrangements is another direction in which a policy for a comprehensive approach could go. This requires detailed standards and specifications for restoration and close inspection and monitoring of operators' activities. This approach is expensive to operate and requires technical inspectors who would make the judgements on breaches of condition. But it is possible to assess failure by reference to the laid down standards and specifications. This is closer to contract compliance procedures than to existing British planning procedures.

Thus, for a comprehensive bonding system to work properly there needs to be:

- either, clearly identified responsibilities for the determination of whether breach of conditions has occurred, with only limited appeal procedures; or

- highly specified schemes of restoration such that technical breach of conditions can be clearly identified.

Both approaches would mark radical departures to British minerals planning procedures. There would be serious difficulties of transition, possibly with two systems running in parallel.

Bonds as a Supplement to Enforcement Procedures

The alternative is to introduce a partial financial guarantee or bonding scheme which would not replace existing enforcement procedures but would supplement them. In practice, this would be achieved only for security against financial or similar failure by the operator (eg. disappearance, imprisonment). Technical failure would remain to be pursued through existing enforcement procedures.

Occurrences of financial failure are clear-cut. They would offer bondsmen no difficulties in terms of release of funds. They would meet some of the concerns of planning authorities. They would not lead to substantial abuse, but safeguards would be difficult to introduce against the strategic bankruptcy.

5.4 Schemes for Financial Guarantees and Bonds

In Chapter 4 two basic approaches to a financial guarantee and bonding scheme were identified, decentralised and national. The decentralised approach would leave the decision as to the financial guarantee or bonding arrangements to the planning authority and the operator. In the national approach there would be a scheme, probably set up under the initiative of Government and possibly involving an institution or agency, which would issue the financial guarantee and to which both the planning authority and operator would refer in the determination of any breach of condition.

Of these two approaches, the decentralised scheme would be easier for Government to implement. But it has a number of limitations, including higher costs, which the Consultants believe should be carefully assessed by Government. A national scheme along the lines of an insurance fund has clear advantages over a national restoration fund type of scheme.

The introduction of a either kind of scheme, decentralised or national, would need to be brought about by a legislative change such as was involved in the granting of powers to the South Wales authorities in their local acts. These would give to the planning authorities the powers to impose conditions which include financial guarantee or bonding arrangements.

Decentralised Scheme

The type of instrument under decentralised schemes should be left to operators and authorities to agree. It is highly desirable that there should be consistency of approach by the planning authorities in relation to the acceptance of particular forms of financial guarantee and bonds. Government guidance would certainly be helpful, even necessary, to the planning authorities concerning different types of security. Without guidance and a consistent approach there is a danger that certain operators will be penalised by conditions imposed upon them. Whilst appeals procedures will be in place to ensure that there is a safeguard for the reasonableness of conditions, Government will still need to form its own view on the security that is being offered by any individual operator when appeals are made.

The advantage of a decentralised scheme is that Government can leave to the planning authorities and the operators the best way to sort matters out. Flexibility is ensured. But a basic disadvantage of this approach is that the costs to the industry and more generally to society are hidden. Evidence from the surveys conducted for this study found that the smaller operators will be at a disadvantage compared with the national operators in securing financial guarantees and bonds. There is a danger that there will be a restraint of trade introduced through this mechanism which will be imposed most directly on the small firm sector.

National Scheme

All operators would need to be members of a national scheme or fund, thereby satisfying the planning authorities and Government as to their financial security. Membership of the scheme would in effect become a licence to operate. The scheme might be operated by a new Agency or could be contracted out. A voluntary scheme is not likely to be feasible.

Stevens argued strongly against a restoration fund approach to a national scheme. Responsibility for restoration would be taken away from operators, with the danger of creeping costs, and there would be the burden of administration. A restoration fund would have substantial funds flow if it covered the minerals industry as a whole.

Far more promising is the concept of an insurance fund which covers only costs associated with the risk of default. Stevens believed that this approach would probably be the best way forward, assuming that a scheme is to be introduced.

The main issue to be decided is the scope of the insurance fund; whether it would cover technical as well as financial failure. The size and operations of the fund would differ greatly, dependent on this choice. If technical failure is included, then the Agency operating the fund could conceivably act in an arbitration capacity in cases of alleged technical default.

An insurance fund would be cheaper to operate than a decentralised scheme because of the principle of insuring a book. However, levies or fees charged to industry may need to the standardised, at least until evidence of default rates was more clear-cut and discriminatory levies could be introduced, which would mean that larger operators would be at a disadvantage in bearing the same costs as smaller operators.

A variant on the national insurance fund would be a national bonding institution which would provide operators with bonds which would be offered against operators' securities.

5.5 Summary of Recommendations

Financial Guarantees and Bonds

The Consultants' principal conclusions and recommendations in relation to financial guarantees and bonds for mineral restoration and aftercare are as follows:

- The Consultants are not in favour of a new scheme of financial guarantees or bonds. However, the basis of many existing voluntary Agreements, which are not common at present but which are likely to increase in the future, is not satisfactory. Future Agreements need to be put onto a consistent basis, in order to avoid discriminatory, costly and essentially irrelevant bonding arrangements coming into effect. This may be achieved by the issue of Minerals Planning Guidance as to the circumstances under which it would be appropriate for a planning authority to reach an Agreement with an operator in relation to financial guarantees, forms of guarantee, checks that the local authority might make on the financial institutions offering guarantees and related matters.

- The Consultants believe that improvements to updating older mineral permissions together with the recent changes to existing enforcement powers are sufficient to achieve restoration objectives in most cases.

- However, if a comprehensive financial guarantee or bonding scheme to cover both financial and technical default was judged to be necessary, then the Consultants recommend that the appropriate route to take would be to keep the existing system of planning consents and to permit financial guarantees and bonds to become conditions of permissions. Legislation would be required. It would be necessary to satisfy bondsmen and others as to the circumstances governing the release of bonds in some cases of default. Most probably the easiest arrangement to introduce would be arbitration procedures to ensure that the release of the financial guarantees and bonds is fair.

- Alternatively, Government could introduce a new scheme which is less ambitious which would provide a partial rather than comprehensive system of financial guarantees and bonds. This limited scheme would be for the purpose of providing security only against the financial or similar failure of an operator to carry out restoration and aftercare following mineral extraction. This partial financial guarantee and bonding system would supplement existing Planning procedures. The limited benefits of guaranteeing against financial failure would need to be weighed against the costs of establishing a new scheme.

- Choices about the approach to guarantees would have to be made in relation to a new scheme, if it is to be introduced. The Consultants believe that a national fund scheme has the advantage over a decentralised scheme, being substantially cheaper for the industry overall and that a national insurance fund would be the best approach. A restoration fund approach is not favoured by the Consultants.

- A national insurance scheme would need to be initiated by Government and would probably need to be operated by a Government Agency although there are other possible options including contracting out. All operators would have to be members of the insurance fund in order to qualify to operate. The levy payable to the fund would need to cover the fund's restoration liabilities, administration and other costs. The levy could be tied to the firms' outputs.

Further Recommendations

Other recommendations as a consequence of undertaking this study, are as follows:

- The Study has not found that failure to make adequate financial provision by operators is a cause of restoration failure. Nevertheless, the tax system for companies acts as a disincentive to make provision in some cases. Government may wish to reconsider tax rules in relation to provision for restoration.

- The possible extent of compensation payments has caused difficulties for authorities wishing to modernise permissions. There is a case for extending IDO procedures to sites that have no restoration conditions at present.

- Borrow pits present authorities with particular enforcement problems. Performance bonding for restoration of borrow pits under the terms of the construction contract might be the best route to ensure compliance in these cases.

APPENDIX 1

Postal Questionnaires of Planning Authorities and Mineral Operators

Summary of Questions used in the Postal Survey of Planning Authorities with Minerals Responsibilities

Degree of Satisfaction

1. In the table below, please enter numbers of sites or part sites for surface mineral working or the surface disposal of mineral working deposits which have ceased or completed operations in the period 1982 to 1990 in your authority's area according to the following categories of restoration and/or aftercare performance. [Note: see report, Table 2.2, for categories and results].

2. Please categorise the number of sites or part sites for surface mineral working or for the surface disposal of mineral working deposits which have ceased or completed operations in the period 1982 to 1990 in your authority's area by type of mineral operator. [Note: see report, Table 2.3, for categories and results].

Reasons for unsatisfactory or no restoration and/or aftercare

3. In respect of those sites identified in questions 1 & 2 where restoration was unsatisfactory or no restoration took place.

    (i) Please indicate how many were due to the following factors. (More than one factor may apply. Please indicate only major ones).

    (a) unrealistic planning conditions

    Please specify problems

    (b) loosely worded planning conditions

    Please specify the nature of these conditions

    (c) no requirement to restore

    (d) the suspension of activities

    (e) the financial failure of the operator

    Please specify the nature of the failure

    (f) shortage of fill materials

    (g) the possibility of reworking the site

    (h) lack of monitoring

    Please specify problems

(i) inadequate practices by mineral operators e.g. soils damaged or lacking

(h) slowness or delays in completing restoration programme not otherwise included above

Please specify these problems

Enforcement

4. In respect of those cases in questions 1 & 2 identified as having unsatisfactory or no restoration

   (i) In how many cases has enforcement action already been taken or will enforcement action be taken to require satisfactory restoration?

   (ii) In how many other cases would it have been possible for the authority to have instituted enforcement action to require satisfactory restoration?

   (iii) Indicate briefly why enforcement action to require satisfactory restoration was either not instituted or has not been effective.

5. In respect of those sites listed as having satisfactory restoration in questions 1 & 2 above.

   (i) In how many cases did you take enforcement action to achieve satisfactory restoration/aftercare

   (ii) In how many cases was a financial safeguard activated to achieve satisfactory restoration/aftercare?

   Please specify the nature of these safeguards

1981 Minerals Act

6. Do you consider that the powers contained in the 1981 Minerals Act have improved the likelihood of satisfactory restoration (including aftercare) in your authority's area?

7. Has your authority undertaken a review of restoration conditions on any permissions or sites following the implementation of the 1981 Act?

   If yes,

   (i) How many permissions or sites have been reviewed?

   (ii) In how many cases has the review suggested there is scope for adding or modifying restoration/aftercare conditions?

   (iii) Have you taken or are your proposing to take any actions to change restoration/aftercare conditions?

If no, why not?

8. Are there any constraints reducing the effectiveness of provisions in the 1981 Minerals Act regarding the review of restoration (including aftercare) conditions?

Please specify

9. If the constraint on the effectiveness of the 1981 Act is financial[1], please indicate whether it is:

   (i) The maximum threshold figure included in the Minerals Compensation regulations?

   (ii) The limitation on abatement of 10% of the notional value of the right to win and work minerals in the land concerned?

   (iii) The limitations on the circumstances in which abatement applies (i.e. satisfying minerals compensation requirements)?

   1   see guidance notes

Monitoring

10. Do you undertake regular monitoring of restoration and aftercare for sites?

11. Who in your planning authority is responsible for monitoring of sites for enforcement/compliance?

    (i) Minerals section personnel

    (ii) Special enforcement section

    (iii) Separate county surveyors department

    (iv) Other sections in the authority

       If yes, please specify

12. Do you produce regular monitoring reports on the condition of mineral sites within your area?

    If no, is this due to:

    (i) a lack of staff resources

    (ii) lack of necessary expertise within the local authority

    (iii) other

       Please specify

Safeguards to ensure the compliance with restoration/aftercare conditions

13. Please supply numbers and additional brief details of Planning permissions granted in the period 1982 to 1990 having safeguards and/or security for the compliance with restoration and/or aftercare conditions by mineral group.

14. Are there any additional comments you wish to make either on the existing systems of Planning control over mineral workings to ensure restoration/aftercare or suggestions for new measures to improve the quality and/or speed of restoration and/or aftercare of mineral workings?

Guidance notes for Mineral Planning Authorities (England & Wales)

Question 1

"Sites for mineral workings" or for the "surface disposal of mineral working deposits" are defined as sites having planning permissions for the winning and working of minerals, in on or under the land and/or for use as a site for the surface disposal or storage of mineral working deposits.

"Mineral types". Where 2 or more minerals are worked or deposited in the same site area, the entry should be made under the major mineral which promoted the application and development.

Clay and Shale - includes ball clay, fireclay and potters clay

Vein Minerals - includes tin, copper, lead, silver, zinc, haematite, iron ore, barytes, calcspar

Whether restoration is "satisfactory" or "unsatisfactory" is in part a subjective judgement on the part of the MPA. However, this judgement should take account of what would be reasonably achievable having regard to the nature of planning conditions, the condition of the site prior to mineral working, the physical constraints of the site, and whether or not the condition of the site is suitable for the intended after use.

"No restoration" refers to sites or part sites where mineral working has ceased or operations completed and no attempt has been made to restore the site.

Question 2

"National Operators" are defined for "common minerals" as having operations in several regions or having a few sites but with a large market share or for specialist minerals (eg. china clay or fluorspar) of national importance.

"Local Operators" are defined as operators not included above; having operations in one or two mpa's or concentrated in one region, and having a small number of sites.

Satisfactory and unsatisfactory restoration is defined as in question 1.

Question 3

Where it is necessary please use additional sheets to give more detail.

Question 4

Enforcement action refers to enforcement undertaken in respect of breaches of planning control or conditions under relevant legislation.

Question 5 (ii)

"Safeguards" refers to mechanisms other than planning conditions such as voluntary agreements under section 106 of the Town and Country Planning Act 1990 (formerly section 52 of the 1971 Act), agreements under S.33 of the Local Government (Miscellaneous Provisions) Act 1982, Other Acts, Local Acts, in England and Wales, or in Scotland Section 50 of the Town and Country Planning Act 1972 or Section 69 of the Local Government Act 1973 and arrangements by contracts. The form of the safeguards may be financial guarantees or bonds, insurance or any other arrangements.

Where relevant or necessary please use additional sheets for your responses.

Questions 6,7,8 & 9

References to the Minerals Act 1981 refers to the Town and Country Planning (Minerals) Act 1981 as embodied in the Town and Country Planning Act 1990 (formerly the Town and Country Planning Act 1971) in England and Wales or Town and Country Planning Act 1972 in Scotland.

Questions 8 & 9

The "review" of mineral permissions refers to the formal duty to undertake such a review in the 1981 Minerals Act as embodied in the Town & Country Planning Act 1990 in England and Wales, and the Town and Country Planning Act 1972 in Scotland and outlined in MPG 4, "The Review of Mineral Working Sites". MPG 4 provides explanations of compensation aspects and SI 1990 Number 803 revised current maximum thresholds.

General

"Restoration condition" as defined by 1981 Minerals Act means a condition attached to a planning permission requiring that after operations for the winning and working of minerals has been completed, the site shall be restored by the use of any or all of the following, namely, subsoil, topsoil and soil-making material.

"Aftercare condition" means a condition attached to a planning permission requiring that such steps shall be taken as may be necessary to bring land to the required standard for whichever of the following uses is specified in the condition, namely

(a)   agriculture

(b)   forestry

(c)   amenity

An aftercare condition may specify

(a)   the steps to be taken

(b)   require that the steps be taken in accordance with an aftercare scheme approved by the mineral planning authority.

An aftercare condition can only be imposed on a permission which is also subject to a restoration condition.

Summary of Questions used in the Postal Survey of Mineral Operators

Background information

1. Please identify the geographical spread of your Mineral sites?

    (i) National   Please tick major operating regions

South West	North West
South East	Yorks and Humber
East Anglia	North
East Midlands	Scotland
West Midlands	Wales

    (ii) Local   Please state local areas

2. In the table below please enter the number of sites or part sites for surface mineral working or the surface disposal of mineral working deposits, controlled by your company, which have ceased or completed operations in the period 1982-1990. [Note: see report, Table 2.7, for categories and results of this question].

Implementing and Monitoring Site Restoration

3. Who in your company is responsible for seeing that planning conditions for restoration and aftercare are carried out?

    (i) quarry managers at individual sites

    (ii) restoration manager covering all sites

    (iii) restoration or estates personnel attached to groups of quarries on a regional/local basis.

    (iv) Others - please specify.

4. If yes to 3 (i) above, what training in restoration do quarry managers receive?

5. What is the policy of your company regarding financial provision for compliance with site restoration and aftercare conditions?

    (i) A general provision is made in advance covering all company sites

(ii) Specific provision is made in advance for each site

(iii) The full provision is made when site development commences

(iv) Provision is accumulated on an incremental basis as sites are developed

(v) No forward provision is made - restoration/aftercare costs are incurred out of current revenue

(vi) Any other approach. Please specify

6. Do you make provision for the restoration of sites which currently have no formal requirements on restoration in their planning permissions?

7. Briefly indicate what form the provision takes

Satisfaction with Restoration/Aftercare Conditions

8. For sites with Planning permissions granted in the period 1982 to 1990 and having restoration and/or aftercare conditions:

   (i) Are there likely to be problems in achieving the standard of restoration/aftercare required by these conditions?

   Please specify the particular problems that are likely to be faced

   (ii) Do you have any sites undergoing formal aftercare as yet?

Restoration and aftercare performance

9. (i) How many of the sites or part sites in the answer to question 2 contained restoration (including aftercare) conditions in their planning permissions

10. Of the sites specified in 9 how many sites if any, were not restored to the planning authority's satisfaction?

   For these sites please indicate how many were due to the following factors. (More than one factor may apply. Please indicate <u>only major ones</u>).

   (i) unrealistic planning conditions

   Please specify problems

   (ii) loosely worded planning conditions

   Please specify the nature of these conditions

   (iii) local authority were not satisfied with the compliance and standard of restoration achieved

Please specify problems

(iv) lack of monitoring and advice from local authorities

Please specify problems

(v) activities were suspended

(vi) the financial costs of the operations

Please specify why you were unable or unwilling to meet these costs

(vii) shortage of fill materials

(viii) the possibility of reworking the site

(viiii) slowness or problems in completing restoration not included above

Please specify problems

11. Of the sites identified in the answer to 10, please indicate the mineral types involved.

Safeguards to ensure the compliance of restoration/aftercare conditions

12. Please give brief details of planning permissions granted in the period 1982 to 1990 having additional safeguards and/or security for the compliance with restoration and/or aftercare conditions.

13. Are there any additional comments you wish to make either on the existing system of planning control over mineral workings or suggestions for new measures to improve the quality and/or speed of restoration and/or aftercare of mineral workings.

Guidance notes for Mineral Operators

Question 1-2

"Mineral sites" are defined as sites or part sites having planning permissions for operations for the winning and working of minerals, in on or under the land; or a site used for the surface disposal and/or storage of mineral working deposits.

Question 2

Mineral types. Where 2 or more minerals are worked or deposited in the same site area, the entry should be made under the major mineral which promoted the application and development.

Clay and Shale - includes ball clay, fireclay and potters clay

Vein Minerals - includes tin, copper, lead, silver, zinc, haematite, iron ore, barytes, calcspar, fluorspar

Question 3

"Restoration condition" means a condition attached to a planning permission requiring that after operations for the winning and working of minerals has been completed, the site shall be restored by the use of any or all of the following, namely subsoil, topsoil and soil making material.

"Aftercare condition" means a condition attached to a planning permission requiring that such steps shall be taken as may be necessary to bring land to the required standard for whichever of the following uses is specified in the condition, namely

(a) agriculture

(b) forestry

(c) amenity

An aftercare condition may specify

(a) the steps to be taken; or

(b) require that the steps be taken in accordance with an aftercare scheme approved by the mineral planning authority.

An aftercare condition can only be imposed on a permission which is also subject to a restoration condition.

Question 4

Training methods to be specified here might include formal training courses at external institutions (please specify institutions and qualifications gained where possible), in house training courses on a regular basis or ad hoc courses.

Question 8

Enforcement action refers to action taken by planning authorities against breaches of planning control or planning conditions.

Question 12

"Additional Safeguards" (or security) for restoration and aftercare refers to mechanisms other than planning conditions such as voluntary agreements under Section 106 of the Town and Country Planning Act (England and Wales) (formerly Section 52 of the 1971 Act) Section 50 of the Town and Country Planning Act (Scotland) 1972, Section 33 of the Local Government (Miscellaneous Provisions) Act 1982 (England and Wales) Section 69 of the Local Government Act 1973 (Scotland). Other Acts, Local Acts or arrangements by contract. The form of safeguards may be financial guarantees or bonds, insurance or any other arrangements.

Where relevant or necessary please use additional sheets for your responses.

General

References to planning authorities means Mineral Planning Authorities in England and Wales and Planning Authorities having responsibility for minerals in Scotland.

APPENDIX 2

Respondents to the Postal Surveys

County Councils

ENGLAND

Bedfordshire County Council
Buckinghamshire County Council
Cambridgeshire County Council
Cheshire County Council
Cornwall County Council
Cumbria County Council
Derbyshire County Council
Devon County Council
Durham County Council
Essex County Council
Gloucestershire County Council
Hampshire County Council
Hertfordshire County Council
Humberside County Council
Isle of Wight County Council
Leicestershire County Council
Lincolnshire County Council
Norfolk County Council
Northamptonshire County Council
Somerset County Council
Suffolk County Council
Surrey County Council
Warwickshire County Council
West Sussex County Council
Wiltshire County Council

WALES

Powys County Council
Gwynedd County Council
Mid Glamorgan County Council
Dyfed County Council

METROPOLITAN DISTRICTS

Greater Manchester

- Bolton Metropolitan Borough Council
- Oldham Metropolitan Borough
- Rochdale Metropolitan Borough
- Stockport Metropolitan Borough
- St Helens Metropolitan Borough
- Wigan Metropolitan Borough

South Yorkshire

- Doncaster Metropolitan Borough Council
- Rotherham Metropolitan Borough

Tyne & Wear

- Gateshead Metropolitan Borough
- North Tyneside Metropolitan Borough
- South Tyneside Metropolitan Borough
- Sunderland Metropolitan Borough

West Midlands

- Dudley Metropolitan Borough
- Solihull Metropolitan Borough
- Walsall Metropolitan

West Yorkshire

- City of Bradford
- Calderdale Metropolitan Borough
- Kirklees Metropolitan Borough
- City of Leeds
- City of Wakefield

National Parks

- Peak National Park
- Pembrokeshire Coast National Park
- Yorkshire Dales National Park

SCOTLAND

Central Regional Council

- Clackmannan District Council
- Falkirk District Council
- Stirling District Council

Fife Regional Council

- Kirkcaldy District Council

Grampian Regional Council

- City of Aberdeen

Lothian Regional Council

- Mid Lothian District Council
- West Lothian District Council

Strathclyde Regional Council

- Bearsden and Milngavie District Council
- Clydesdale District Council
- Cumnock and Doon Valley District Council
- Cunninghame District Council
- East Kilbride District Council
- Eastwood District Council
- Glasgow District Council
- Kilmarnock and Londonn District Council
- Hamilton District Council
- Motherwell District Council
- Strathkelvin District Council

Tayside Regional Council

- City of Dundee District Council

Perth and Kinross District Council

Orkney Islands Council

Western Isles Islands Council

Mineral Operators

Alresford S&B Co Ltd
ARC Group
Bain Aggregates Ltd
Barker Bros Aggregates
Blue Circle Industries Plc
Bonnar Sand & Gravel Co Ltd
Bromfield Sand and Gravel Co
Bruntingthorpe Gravels Ltd
Bulbricks Co. Ltd
Butterley Brick Ltd
Coal Contractors Ltd
Cory Environmental Aggregates
Cults Lime Ltd
Dimensional Stone Ltd
Drinkwater Sabey Ltd
ECC International Ltd
Ennemix Aggregates Ltd
Euston Lime Co Ltd
Evered Bardon (England) Ltd
Fife Silica Sands Ltd
Fisons Plc. (Horticulture Division)
Foster Yeomen Ltd
Glendinning Group of Companies
E & JW Glendinning Ltd
Greenham Construction Materials Ltd
H Tuckwell and Sons Ltd
Hall Aggregates (Eastern Counties)
Hall Aggregates (South Coast) Ltd
Hall Aggregates (South East) Ltd
Hardrock Ltd
Hargreaves Quarries Ltd
Harleyford Aggregates Ltd
Hills Aggregates Ltd
Holderness Aggregates Ltd
HSS Engineering Ltd
Huntsmans Quarries Ltd
J & B Martin (C & F) Ltd
John Bourne and Co
John Wainwright & Co Ltd
John Fyfe Ltd
Joseph Arnold & Sons Ltd
Kirkstone Quarries Ltd
Lander's Quarries
Laporte Minerals

Mansfield Sand Company Ltd
Marley Roof Tile Co Ltd
McAlpine Quarry Products
McLaren Roadstone Ltd
Meldon Quarry Ltd
Moreton C. Cullimore (Gravels) Ltd
Multi-Agg Ltd
Nash Rocks Ltd
Needham Chalks Ltd
New Milton Sand and Gravel Co
Nickolls Quarries Ltd
North West Aggregates Ltd
Northern Aggregates (Subsid RMC Plc)
Parsons (Norwood) Ltd
Pattersons of Greenoakhill Ltd
Pioneer Aggregates UK Ltd
R & D Aggregates
R & A Young Mining Ltd
RMC Roadstone Ltd/RMC Ind Mins
Roger Constant & Co Ltd
Salop Sand & Gravel Supply Co Ltd
Sandsfield Gravel Co Ltd
Scottish Aggregates
Singleton Birch Ltd
Smith & Sons (Bletchington) Ltd
St Albans Sand & Gravel Co Ltd
Steetley Quarry products
Stonegrave Aggregates Ltd
Summerleaze Ltd
Tarmac Roadstone Ltd
Tilcon Ltd
Tinto Sand and Gravel Ltd
Tarmac Roadstone Ltd (Western)
Thos W Ward Roadstone Ltd
Tillicoultry Quarries Ltd
United Fireclay Products Ltd
United Glass Ltd
W M White Plc
John Wainwright & Co Ltd
Wanlip Gravels Ltd
Watts Blake Bearne & Co Plc
Western Roadstone
Western Aggregates
William Boyer & Sons Ltd
Wimpey Hobbs Ltd
Woodhall Spa S&G Co Ltd
Wotton Roadstone Ltd
WW Graham (Contractors) Ltd

APPENDIX 3

Participants in the Follow-Up Surveys

MPAs

Authority	Personnel
Bedfordshire	Ms R Pillar
Cornwall	Mark Jones
Cumbria	Lucy Binnie
Derbyshire	Roger Caisley Chris Drury
Devon	Stuart Redding David Prossley
Doncaster	Chris Green Arthur Doyle
Dyfed	Peter Kendall Carol Lawrence
Highland Regional Council	John Greaves Andrew Brown
Hampshire	Richard Reed Roger Stow
Leicestershire	Craig Ball
Mid Glamorgan	Michael Gandy Ruth Amundson
Northumberland County Council	Gordon Halliday Andrew Brack
Peak District National Park	Howard White
Shetland Islands	Mr Pert
Somerset	Richard Moon Ken Hobden Helen Wass

Surrey	Stephen Lawrenson
Wakefield	Michael Elliot Martin Seddon
West Lothian	Karen Patterson
Wigan	Dennis McBride

Mineral Operators

Company	Personnel
ARC	David Thomas
BACMI	Duncan Pollock
Blue Circle	Michael Bellingham
British Coal (Opencast)	John Kanefsky John Booth
Butterley Brick	Nicholas Hamilton
Foster Yeoman	David Tidmarsh Hugh Lucas
Greenham Construction	Richard Millican
Laporte Minerals	John Bramley
Mansfield Sand and Gravel	Mr Abrahams
Multi-Agg Ltd	Cliff Puffet
Redland Aggregates	Stephen Savery John Leivers Charles Peachey Chris Robson
RMC (Roadstone)	Colin D'Oyley
RMC (Feltham)	Bryan Frost John Kimberley
SAGA	Terry MacIntyre

Summerleaze   Mr Kirkpatrick

Tarmac   Ron Parry

Watts Blake & Bearne   John Briggs

Wimpey Hobbs   Anne Dugdale

Financial Institutions

American International Underwriters   Michael Anderson

DoE Working Group on Financial Guarantees Against
  Environmental Damage from Landfill Sites   Michael Burns

Fenchurch Insurance Brokers   Michael Stott
                                                     John Dunn

Hambros Bank   Matthew Brooks
                                                     David Chubb

Hogg Insurance Brokers   Nigel Allington

Lloyds Bank   Anthony Woodward

National Westminister Bank   Ian Scurfield
                                                     Frank Scott
                                                     Simon Mockford

Sedgwick   Douglas Masham
                                                     Michael Rutherford
                                                     Adam Golder

Sun Alliance Financial Risks   Fiona Willis

Representatives of larger mineral operators interviewed for bonding perspective

British Coal Opencast	John Kanefsky
	John Booth
Foster Yeoman	David Tidmarsh
	Hugh Lucas
RMC Group	Bryan Frost
	John Kimberley
Redlands	Stephen Savery
	John Leivers
Tarmac	Ron Parry